Problems and Methods in Analysis

VOLUME 2

Problems and Methods in Analysis

VOLUME 2

by

W. KRYSICKI; L. WŁODARSKI
A. J. ZIELICKI; D. KONSTANT

PERGAMON PRESS
OXFORD · LONDON · EDINBURGH · NEW YORK
TORONTO · PARIS · BRAUNSCHWEIG

Pergamon Press Ltd., Headington Hill Hall, Oxford
4 & 5 Fitzroy Square, London W.1
Pergamon Press (Scotland) Ltd., 2 & 3 Teviot Place, Edinburgh 1
Pergamon Press Inc., 44-01 21st Street, Long Island City, New York 11101
Pergamon of Canada, Ltd., 6 Adelaide Street East, Toronto, Ontario,
Pergamon Press S.A.R.L., 24 rue des Écoles, Paris 5e
Friedr. Vieweg & Sohn Verlag, Postfach 185, 33 Braunschweig,
West Germany

First edition 1966
Library of Congress Catalog Card No. 65-29065

This book is based on an original Polish work by
W. Krysicki and L. Włodarski entitled *Analiza Matematyczna w Zadaniach* published in two volumes
by Panstwowe Wydawnictwo Naukowe, Warsaw, 1958.

(2744/66)

CONTENTS

FOREWORD

This book is based on a two volume book by W. Krysicki and L. Włodarski published in Warsaw in 1952 under the title *Analiza Matematyczna w Zadaniach.* It has been our aim to adapt the work so that it is suitable for sixth form students, particularly scholarship students, and for those first year university students (not necessarily mathematical specialists) who need a systematic course of revision in the methods of calculus. To this end a considerable amount has been omitted from the original (notably a full treatment of differential equations), and various sections have been added to or completely rewritten.

The purpose of the book is to help the student solve problems in analytical calculus without allowing him to forget the theory involved or to approach the matter in too mechanical a way. Consequently each chapter begins with a statement without proof of any necessary definitions and theorems. There is then a set of fully worked examples, followed by a set of exercises. Answers to all the exercises, together with hints to the solution of some of the harder problems, are given at the end of the book.

It must be insisted that this is not a textbook, but should be used in conjunction with the textbook of the student's choice. We hope that carefully used in this way the book will be of real assistance to the student in helping him to master the niceties of analysis.

Our thanks must go to Professor W. Krysicki and Dr L. Włodarski, for allowing us so free a hand with their material, to Dr E. A. Maxwell for his encouragement, and to the Pergamon Press for their care and efficiency.

D. K. and A. J. Z.

Cardinal Vaughan School, 1964.

CHAPTER 10

INDEFINITE INTEGRALS

§ 1. Introduction

Given a function $f(x)$ defined in the interval $a<x<b$, there may be associated with it a function $F(x)$, called its primitive function, defined in the same region, such that $F'(x) = f(x)$. Two functions having in this interval the same finite first derivative can differ only by a constant. Thus functions whose first derivative is $2x$ are x^2+3, x^2+5, or in general, x^2+C.

The indefinite integral of $f(x)$ with respect to x, written

$$\int f(x)\,\mathrm{d}x$$

is $F(x)+C$, where $F(x)$ is the primitive function of $f(x)$ and C is an arbitrary constant.

That is,

$$\int f(x)\,\mathrm{d}x = F(x)+C,$$

where

$$F'(x) = f(x).$$

§ 2. Some standard forms

2.1 $$\int x^n\,\mathrm{d}x = \frac{x^{n+1}}{n+1}+C \qquad (n \neq -1,\ x > 0).$$

If n is a natural number the condition $x > 0$ is not required. If n is a negative integer then it is sufficient to ensure that

$x \neq 0$. Three examples are given below:

(a) $n = 0 \qquad \int dx = x + C;$

(b) $n = -\frac{1}{2} \qquad \int \frac{dx}{\sqrt{x}} = 2\sqrt{x} + C \qquad (x > 0);$

(c) $n = -2 \qquad \int \frac{dx}{x^2} = -\frac{1}{x} + C \qquad (x \neq 0).$

2.2 $\int \frac{dx}{x} = \log|x| + C \qquad (x \neq 0).$

Note that $\log z$ is defined only for $z > 0$.

2.3 $\int e^x \, dx = e^x + C.$

2.4 $\int a^x \, dx = \frac{a^x}{\log a} + C \qquad (a > 0,\ a \neq 1).$

2.5 $\int \cos x \, dx = \sin x + C.$

2.6 $\int \sin x \, dx = -\cos x + C.$

2.7 $\int \sec^2 x \, dx = \tan x + C \qquad (\cos x \neq 0).$

2.8 $\int \operatorname{cosec}^2 x \, dx = -\cot x + C \qquad (\sin x \neq 0).$

2.9 $\int \frac{dx}{\sqrt{(1-x^2)}} = \text{arc sin } x + C = -\text{arc cos } x + C'^{*}$

$(|x| < 1).$

2.10 $\int \frac{dx}{\sqrt{(a^2-x^2)}} = \text{arc sin } \frac{x}{a} + C = -\text{arc cos } \frac{x}{a} + C'$ *

$(x^2 < a^2).$

2.11 $\int \frac{dx}{1+x^2} = \text{arc tan } x + C = -\text{arc cot } x + C'.$*

2.12 $\int \frac{dx}{a^2+x^2} = \frac{1}{a} \text{arc tan } \frac{x}{a} + C = -\frac{1}{a} \text{arc cot } \frac{x}{a} + C'.$*

2.13 $\int \sinh x \, dx = \cosh x + C.$

2.14 $\int \cosh x \, dx = \sinh x + C.$

2.15 $\int \frac{dx}{\cosh^2 x} = \tanh x + C.$

2.16 $\int \frac{dx}{\sinh^2 x} = -\coth x + C.$

2.17 $\int \frac{dx}{\sqrt{(x^2+1)}} = \sinh^{-1} x + C = \log [x + \sqrt{(x^2+1)}] + C.$

[*Note:* The inverse hyperbolic functions are denoted in many books by ar sinh, ar cosh etc., where ar is an abbreviation for area.]

2.18 $\int \frac{dx}{\sqrt{(x^2+a^2)}} =$

$$= \sinh^{-1} \frac{x}{a} + C = \log \left[\frac{x}{a} + \frac{1}{a} \sqrt{(x^2+a^2)} \right] + C.$$

2.19 $\int \frac{dx}{\sqrt{(x^2-1)}} = \cosh^{-1} x + C = \log |x + \sqrt{(x^2-1)}| + C$

$(|x| > 1).$

---

* Note that $\text{arc sin } z + \text{arc cos } z = \frac{\pi}{2}$,

$\text{arc tan } z + \text{arc cot } z = \frac{\pi}{2}$,

and $\text{arc sec } z + \text{arc cosec } z = \frac{\pi}{2}$.

2.20 $$\int \frac{dx}{\sqrt{(x^2-a^2)}} = \cosh^{-1}\frac{x}{a}+C = \log\left|\frac{x}{a}+\frac{1}{a}\sqrt{(x^2-a^2)}\right|+C \quad (x^2 > a^2).$$

2.21 $$\int \frac{dx}{1-x^2} = \tanh^{-1} x+C = \frac{1}{2}\log\left|\frac{1+x}{1-x}\right|+C \quad (x^2<1).$$

2.22 $$\int \frac{dx}{a^2-x^2} = \frac{1}{a}\tanh^{-1}\frac{x}{a}+C = \frac{1}{2a}\log\left|\frac{a+x}{a-x}\right|+C \quad (x^2 < a^2).$$

2.23 $$\int \frac{dx}{x^2-1} = -\coth^{-1}x+C = \frac{1}{2}\log\left|\frac{x-1}{x+1}\right|+C \quad (x^2 > 1).$$

2.24 $$\int \frac{dx}{x^2-a^2} = -\frac{1}{a}\coth^{-1}\frac{x}{a}+C = \frac{1}{2a}\log\left|\frac{x-a}{x+a}\right|+C \quad (x^2 > a^2).$$

2.25 $$\int \frac{dx}{x\sqrt{(1-x^2)}} = \pm\,\mathrm{sech}^{-1}\,x+C \quad (x^2 < 1).$$

2.26 $$\int \frac{dx}{x\sqrt{(a^2-x^2)}} = \pm\frac{1}{a}\,\mathrm{sech}^{-1}\frac{x}{a}+C \quad (x^2 < a^2).$$

2.27 $$\int \frac{dx}{x\sqrt{(x^2+1)}} = \pm\,\mathrm{cosech}^{-1}\,x+C.$$

2.28 $$\int \frac{dx}{x\sqrt{(x^2+a^2)}} = \pm\frac{1}{a}\,\mathrm{cosech}^{-1}\frac{x}{a}+C.$$

§ 3. Some general properties of indefinite integrals

3.1 The addition of integrals

$$\int \{f(x)+g(x)\}\,dx = \int f(x)\,dx+\int g(x)\,dx.$$

3.2 Multiplication by a constant

$$\int k\,f(x)\,\mathrm{d}x = k\int f(x)\,\mathrm{d}x.$$

3.3 Integration by parts

If u and v are functions of x with continuous first derivatives, then

$$\int u\,\mathrm{d}v = uv - \int v\,\mathrm{d}u.$$

3.4 Substitution

Given a function $g(x) = u$, then $g'(x)\,\mathrm{d}x = \mathrm{d}u$, and

$$\int f\{g(x)\}\,g'(x)\,\mathrm{d}x = \int f(u)\,\mathrm{d}u.$$

3.5 Two important results

(a) $$\int \frac{f'(x)}{f(x)}\,\mathrm{d}x = \log f(x) + C,$$

(b) $$\int \frac{f'(x)}{\sqrt{[f(x)]}}\,\mathrm{d}x = 2\sqrt{[f(x)]} + C.$$

Examples

1. Integrate

$$I = \int x(x-1)(x-2)\,\mathrm{d}x.$$

Solution. The integral can be expressed as the sum of integrals by multiplying out.

$$I = \int (x^3 - 3x^2 + 2x)\,\mathrm{d}x,$$

$$= \int x^3\, dx - 3\int x^2\, dx + 2\int x\, dx,$$

$$= \frac{x^4}{4} - 3\cdot\frac{x^3}{3} + 2\cdot\frac{x^2}{2} + C.$$

Hence,
$$I = \frac{1}{4}x^4 - x^3 + x^2 + C.$$

2. A point moves in a straight line so that its acceleration a at time t is given by:

$$a = 12t^2 + 18 \sin 3t - 2.$$

Find its velocity v and its distance from the origin x at time t if at $t = 0$, $v_0 = 10$ and $x_0 = 5$.

Solution

(i)
$$v = \int a\, dt = \int (12t^2 + 18 \sin 3t - 2)\, dt,$$

$$= 4t^3 - 6 \cos 3t - 2t + C.$$

For $t = 0$,

$$v_0 = -6 + C = 10, \quad \text{hence} \quad C = 16.$$

Hence,

$$v = 4t^3 - 6 \cos 3t - 2t + 16.$$

(ii)
$$x = \int v\, dt = \int (4t^3 - 6 \cos 3t - 2t + 16)\, dt,$$

$$= t^4 - 2 \sin 3t - t^2 + 16t + C'.$$

For $t = 0$,

$$x_0 = C' = 5.$$

Hence,

$$x = t^4 - 2 \sin 3t - t^2 + 16t + 5.$$

3. Integrate

$$I = \int \frac{\sqrt{x} - \sqrt[3]{x}}{x^2}\, dx \qquad (x > 0).$$

Solution. Express the roots of x as fractional powers. Thus

$$I=\int\frac{x^{1/2}-x^{1/3}}{x^2}\,\mathrm{d}x,$$

$$=\int(x^{1/2-2}-x^{1/3-2})\,\mathrm{d}x,$$

$$=\int(x^{-3/2}-x^{-5/3})\,\mathrm{d}x,$$

$$=\frac{x^{-1/2}}{-\frac{1}{2}}-\frac{x^{-2/3}}{-\frac{2}{3}}+C.$$

Hence, $$I=\frac{3}{2\sqrt[3]{x^2}}-\frac{2}{\sqrt{x}}+C.$$

4. Integrate

$$I=\int\frac{x\,\mathrm{d}x}{(x^2+a^2)^n}\qquad(a\neq 0).$$

Solution. Make the substitution $x^2+a^2=u$, then $x\,\mathrm{d}x=(1/2)\,\mathrm{d}u$.

Hence, $$I=\frac{1}{2}\int\frac{\mathrm{d}u}{u^n}.$$

There are two solutions:

(i) $n\neq 1$.

Then, $$I=\frac{1}{2}\,\frac{u^{-n+1}}{-n+1}+C,$$

$$=\frac{-1}{2(n-1)\,(x^2+a^2)^{n-1}}+C.$$

(ii) $n=1$.

Then, $$I=\frac{1}{2}\log u+C,$$

$$=\frac{1}{2}\log(x^2+a^2)+C.$$

5. Integrate

$$I=\int\frac{\mathrm{d}x}{\sqrt{(2x-3)}}\quad\left(x>\frac{3}{2}\right).$$

Solution. Make the substitution $2x-3 = t^2$ (where $t > 0$), then

$$dx = t\,dt.$$

Hence,

$$I = \int \frac{t\,dt}{t} = \int dt = t+C$$

or in terms of x

$$I = \sqrt{(2x-3)}+C.$$

6. Integrate

$$I = \int x\,e^{x^2}\,dx.$$

Solution. Make the substitution $x^2 = t$, then $x\,dx = \frac{1}{2}\,dt$.

Hence,

$$I = \frac{1}{2}\int e^t\,dt = \frac{1}{2}\,e^t+C$$

and

$$I = \frac{1}{2}\,e^{x^2}+C.$$

$\left[\text{Note that } \int e^{x^2}\,dx \text{ is not capable of such simple treatment.}\right]$

7. Integrate

$$I = \int \sin x \cos x\,dx.$$

Solution. Three possible methods are given.

(i) Put $\sin x = t$, then $\cos x\,dx = dt$, and

$$I = \int t\,dt = \frac{1}{2}\,t^2+C,$$

$$= \frac{1}{2}\sin^2 x+C.$$

(ii) Write $\sin x \cos x$ as $\dfrac{1}{2}\sin 2x$, then

$$I = \frac{1}{2}\int \sin 2x \, dx = \frac{1}{4}\int \sin 2x \, d(2x),$$

$$= -\frac{1}{4}\cos 2x + C'.$$

(iii) Put $\cos x = t$, then $\sin x \, dx = -dt$, and

$$I = -\int t \, dt = -\frac{1}{2}t^2 + C'',$$

$$= -\frac{1}{2}\cos^2 x + C'',$$

It may appear that we have three different solutions, but in fact the three solutions differ from each other only by a constant.

$$\frac{1}{2}\sin^2 x - \left(-\frac{1}{4}\cos 2x\right) = \frac{1}{2}\sin^2 x + \frac{1}{4}\cos 2x,$$

$$= \frac{1}{4}(2\sin^2 x + \cos 2x),$$

$$= \frac{1}{4}(2\sin^2 x + 1 - 2\sin^2 x),$$

$$= \frac{1}{4}.$$

That is $C = C' + \dfrac{1}{4}$.

Similarly it can be shown that $C = C'' + \dfrac{1}{2}$.

8. Integrate

$$I = \int x e^x \, dx.$$

Solution. Integrating by parts we put

$$u = x, \quad \text{and} \quad dv = e^x\,dx,$$

then

$$du = dx \quad \text{and} \quad v = e^x.$$

Hence,

$$I = xe^x - \int e^x\,dx, \qquad \text{(by 3.3)}$$

$$= xe^x - e^x + C,$$

$$= e^x(x-1) + C.$$

9. Integrate

$$I = \int \log x\,dx \quad (x > 0).$$

Solution. Integrating by parts we put

$$u = \log x \quad \text{and} \quad dv = dx,$$

then

$$du = \frac{1}{x}\,dx, \quad \text{and} \quad v = x.$$

Hence,

$$I = x \log x - \int x \cdot \frac{dx}{x},$$

$$= x \log x - x + C,$$

$$= x(\log x - 1) + C.$$

10. Integrate

$$I = \int e^x \sin x\,dx.$$

Solution. Integrating by parts we put

$$u = \sin x, \quad \text{and} \quad dv = e^x dx,$$

then

$$du = \cos x\,dx, \quad \text{and} \quad v = e^x.$$

Hence,

$$I = e^x \sin x - \int e^x \cos x\,dx.$$

Similarly, by integrating $\int e^x \cos x \, dx$ by parts we obtain

$$\int e^x \cos x \, dx = e^x \cos x + \int e^x \sin x \, dx.$$

Hence,

$$I = e^x \sin x - e^x \cos x - \int e^x \sin x \, dx,$$

$$= e^x \sin x - e^x \cos x - I,$$

thus,

$$I = \frac{1}{2} e^x (\sin x - \cos x) + C.$$

EXERCISE 18

Integrate the following functions of x with respect to x, stating where necessary any limits on the value of x:

1. $\int \left(5x^2 - 6x + 3 - \frac{2}{x} + \frac{5}{x^2}\right) dx.$ **2.** $\int \frac{(x^2-1)^3}{x} \, dx.$

3. $\int (x^2 - x + 1)(x^2 + x + 1) \, dx.$ **4.** $\int (x^2+4)^5 x \, dx.$

5. $\int \frac{x \, dx}{1+x^2}.$ **6.** $\int \frac{x \, dx}{(x^2+3)^6}.$ **7.** $\int \frac{x^2 \, dx}{a^3+x^3}.$

8. $\int \frac{x \sqrt[3]{x} + \sqrt[4]{x}}{x^2} \, dx.$ **9.** $\int \frac{x \sqrt{x} - x \sqrt[4]{x}}{\sqrt{x}} \, dx.$

10. $\int (3 + 2 \sqrt[4]{x})^3 \, dx.$ **11.** $\int \frac{\sqrt{x} - 2 \sqrt[3]{x^2} + 4 \sqrt[4]{(5x^3)}}{6 \sqrt[3]{x}} \, dx.$

12. $\int \frac{3 + 5 \sqrt[3]{x^2}}{\sqrt{x^3}} \, dx.$ **13.** $\int \sqrt{(3x+1)} \, dx.$

14. $\int \sqrt{(a+bx)} \, dx.$ **15.** $\int \frac{x \, dx}{\sqrt[3]{(2x^2-1)}}.$

16. $\int x \sqrt{(1+x^2)} \, dx.$ **17.** $\int \frac{x}{\sqrt{(3-5x^2)}} \, dx.$

18. $\int \frac{x-1}{\sqrt[3]{(x+1)}}\,dx.$ **19.** $\int \frac{x}{\sqrt{(x^2-6)}}\,dx.$

20. $\int \frac{x^2\,dx}{\sqrt[5]{(x^3+1)}}.$ **21.** $\int \frac{e^{1/x}dx}{x^2}.$ **22.** $\int xe^{-x^2}\,dx.$

23. $\int \frac{dx}{2\cos^2(3x)}.$

24. $\int x\sin(2x^2+1)\,dx.$ **25.** $\int \sin^5 x\cos x\,dx.$

26. $\int \frac{\cos x}{\sqrt{(1+\sin x)}}\,dx.$ **27.** $\int \frac{\sin x}{a+b\cos x}\,dx.$

28. $\int \cos x\exp(\sin x)\,dx$ (where $\exp(z)$ is e^z).

29. $\int x^3\sec^2(x^4)\,dx.$ **30.** $\int \sec^2 x\tan^2 x\,dx.$

31. $\int \frac{x^2\,dx}{\cos^2(x^3+1)}.$ **32.** $\int \frac{(\log x)^2}{x}\,dx.$

33. $\int \frac{dx}{e^x+e^{-x}}.$ **34.** $\int \frac{e^x\,dx}{2e^x+1}.$ **35.** $\int x\log(1+x^2)\,dx.$

36. $\int \frac{\sqrt{(2+\log|x|)}}{x}\,dx.$ **37.** $\int 6^{1-x}\,dx.$

38. $\frac{dx}{x\sqrt{[1-(\log|x|)^2]}}.$ **39.** $\int \frac{\log|\arctan x|}{1+x^2}\,dx.$

40. $\int xe^{x^2}(x^2+1)\,dx.$ **41.** $\int \frac{x^2\,dx}{\sqrt{(1-x^6)}}.$

42. $\int \frac{dx}{(1+x^2)\arctan x}.$ **43.** $\int \frac{(\pi-\arcsin x)\,dx}{\sqrt{(1-x^2)}}.$

44. $\int \frac{x\,dx}{x^4+1}.$ **45.** $\int x^4(1+x)^3\,dx.$ **46.** $\int x^2e^x\,dx.$

47. $\int x^3e^x\,dx.$ **48.** $\int x^4e^{2x}\,dx.$ **49.** $\int x\cos x\,dx.$

50. $\int x^2\cos x\,dx.$ **51.** $\int x^2\sin 5x\,dx.$ **52.** $\int e^x\cos x\,dx.$

53. $\int e^{-2x}\sin 3x\,dx.$ **54.** $\int e^{-x}\cos\frac{2}{3}x\,dx.$ **55.** $\int \sqrt{x}\log x\,dx.$

56. $\int (\log|x|)^3\,dx.$ **57.** $\int \frac{(\log|x|)^2}{x^5}\,dx.$

58. $\int \sqrt{x}\,(\log|x|)^3\,dx.$ **59.** $\int \frac{\log|x|}{x^4}\,dx.$

60. $\int \frac{(\log x)^2}{\sqrt{x}}\,dx.$ **61.** $\int x^3 (\log x)^3\,dx.$

62. $\int x^n \log x\,dx, \quad n \neq -1.$

§ 4. The integration of rational functions

4.1 Introduction

The rational function $f(x)$ is the ratio of two polynomials,

i.e. $$f(x) \equiv \frac{P(x)}{Q(x)} \equiv \frac{a_p x^p + a_{p-1}x^{p-1} + \ldots + a_0}{b_q x^q + b_{q-1}x^{q-1} + \ldots + b_0}.$$

If $p \geqslant q$ the numerator may be divided by the denominator so as to obtain a polynomial of order $p-q$ (which is readily integrable, see 3.1), and a rational function $R(x)/Q(x)$ in which the order of $R(x)$ is less than that of $Q(x)$.

If $p < q$ it can be shown that $f(x)$ can be expressed as the sum of terms like

$$(1) \quad \frac{A}{(hx+k)^m}, \quad \text{and} \quad (2) \quad \frac{\alpha x + \beta}{(ax^2+bx+c)^n},$$

where A, α, β, h, k, a, b, c are constants and m and n are positive integers. Also it will follow that ax^2+bx+c cannot be further resolved into real factors, since if this were possible the terms like (2) could be expressed as the sums of terms like (1). This implies that the discriminant of the quadratic expression is less than zero, i.e. $b^2-4ac < 0$. In general then,

$$f(x) \equiv \sum_{r=1}^{m} \frac{A_r}{(h_1 x + k_1)^r} + \ldots + \sum_{r=1}^{s} \frac{J_r}{(h_j x + k_j)^r}$$
$$+ \sum_{r=1}^{n} \frac{\alpha_r x + \beta_r}{(a_1 x^2 + b_1 x + c_1)^r} + \ldots + \sum_{r=1}^{t} \frac{\lambda_r x + \mu_r}{(a_\lambda x^2 + b_\lambda x + c_\lambda)^r}.$$

Note that some of the coefficients A, J, α, β, λ, μ may be zero.

In expressing $f(x)$ in this way we are said to resolve $f(x)$ into its partial fractions. The examples below (4.2) will make the technique clear.

Examples

1. Integrate

$$I = \int \frac{3x+4}{x-1}\,\mathrm{d}x.$$

Solution. Note that the integral does not exist at $x = 1$. The two polynomials are of the same order, so dividing numerator by denominator we obtain

$$\frac{3x+4}{x-1} = 3 + \frac{7}{x-1}.$$

Hence,
$$I = 3\int \mathrm{d}x + 7\int \frac{\mathrm{d}x}{x-1},$$
$$= 3x + 7\log|x-1| + C.$$

2. Integrate

$$I = \int \frac{\mathrm{d}x}{ax+b}. \quad \left(x \neq -\frac{b}{a}\right).$$

Solution. Put $ax+b = t$, then $\mathrm{d}x = \frac{1}{a}\mathrm{d}t$.

Hence,

$$I = \frac{1}{a}\int \frac{\mathrm{d}t}{t} = \frac{1}{a}\log|t| + C,$$
$$= \frac{1}{a}\log|ax+b| + C.$$

3. Integrate

$$I = \int (ax+b)^n\,\mathrm{d}x.$$

Solution. The values of x for which this function is integrable depend on the value of n.

(i) n a positive integer: no restriction on x;

(ii) n a negative integer: $ax+b \neq 0$;

(iii) n a fraction $\frac{p}{q}$, q odd: no restriction on x;

(iv) n a fraction $\frac{p}{q}$, q even: $ax+b > 0$.

The case $n = -1$ has been considered in the example above.

As before put $ax+b = t$, then $\mathrm{d}x = \frac{1}{a}\,\mathrm{d}t$.

Hence,

$$I = \frac{1}{a}\int t^n\,\mathrm{d}t = \frac{1}{a}\,\frac{t^{n+1}}{n+1} + C,$$

$$= \frac{(ax+b)^{n+1}}{a(n+1)} + C \quad (n \neq -1).$$

4.2 Integrals of the type

$$\int \frac{mx+n}{ax^2+bx+c}\,\mathrm{d}x.$$

(i) If $mx+n$ is the derivative of ax^2+bx+c the integral is a logarithmic function (see 2.21), i.e.

$$\int \frac{2ax+b}{ax^2+bx+c}\,\mathrm{d}x = \log|ax^2+bx+c| + C.$$

We must note that the function will not be integrable for values of x for which the denominator vanishes. The discriminant $\Delta \equiv b^2-4ac$ should be examined to see whether real factors exist. If $\Delta < 0$ the quadratic expression has no real factors and the integral exists for all values of x; if $\Delta \geqslant 0$ the quadratic has real factors and there are either two values or one double value of x (when $\Delta = 0$) for which the integral does not exist.

Examples

4. A point moves in a straight line so that its acceleration a after time t ($t > 0$) is given by

$$a = 6t + \frac{1}{t+1}.$$

Find its velocity v and its distance s from the origin at time t given that at $t = 0$, $v = v_0$ and $s = s_0$.

Solution

$$v = \int a\,dt = \int \left(6t + \frac{1}{t+1}\right) dt,$$

$$= 3t^2 + \log|t+1| + C.$$

Hence, $v_0 = 0 + 0 + C$,

and $v = 3t^2 + \log|t+1| + v_0.$

$$s = \int v\,dt = \int (3t^2 + \log|t+1| + v_0)\,dt,$$

$$= t^3 + (t+1)\log|t+1| - t + v_0 t + C'.$$

Hence, $s_0 = 0 + 0 - 0 + 0 + C'$,

and $s = t^3 + (t+1)\log|t+1| - t + v_0 t + s_0.$

(Note here that t is essentially positive and that the denominator $(t+1)$ can therefore never vanish.)

5. Integrate

$$I = \int \frac{6x-1}{3x^2 - x + 2}\,dx.$$

Solution. We first examine the sign of the discriminant,

$$\Delta = 1 - 24 < 0.$$

Hence the quadratic expression has no real factors and the function is integrable for all real values of x.

Moreover $\frac{d}{dx}(3x^2 - x + 2) = 6x - 1,$

i.e. the numerator is the exact derivative of the denominator,

hence, $$I = \log(3x^2 - x + 2) + C.$$

6. Integrate

$$I = \int \frac{x-3}{x^2-6x+5}\,dx.$$

Solution. Here $\Delta = 36-20 > 0$, hence the quadratic expression factorizes. In fact

$$x^2-6x+5 = (x-5)(x-1).$$

Thus the function is not integrable for $x = 5$ or $x = 1$.

Also $$\frac{d}{dx}(x^2-6x+5) = 2x-6 = 2(x-3).$$

Hence, $$I = \frac{1}{2}\int \frac{2x-6}{x^2-6x+5}\,dx = \frac{1}{2}\log|x^2-6x+5|+C.$$

[Note that if $\Delta \geqslant 0$ this method is useful provided only that the numerator is a multiple of the exact derivative of the denominator.]

(ii) If $mx+n$ is not the derivative of ax^2+bx+c the method of integration will depend on the sign of the discriminant Δ.

(1) $\Delta > 0$. The quadratic can therefore be resolved into real (but not necessarily national) factors and the function split into a pair of partial fractions which can then be integrated as shown above.

Examples

7. Integrate

$$I = \int \frac{dx}{2x^2+5x+3}.$$

Solution. Examining the sign of the discriminant we see that

$$\Delta = 25-24 > 0.$$

By inspection

$$2x^2+5x+3 = (2x+3)(x+1).$$

Hence the function is not integrable for $x = -1$ or $x = -3/2$. Suppose

$$\frac{1}{2x^2+5x+3} \equiv \frac{A}{2x+3} + \frac{B}{x+1}.$$

Then, $$1 \equiv A(x+1) + B(2x+3).$$

Equating coefficients of the powers of x we obtain

$$A+2B = 0, \quad \text{and} \quad A+3B = 1,$$

hence, $$A = -2 \quad \text{and} \quad B = 1.$$

Therefore,

$$I = \int \left\{\frac{1}{x+1} - \frac{2}{2x+3}\right\} dx,$$

$$= \log|x+1| - \log|2x+3| + C.$$

Hence, $$I = \log\left|\frac{x+1}{2x+3}\right| + C.$$

8. Integrate

$$I = \int \frac{11x-1}{3x^2-5x-2}\, dx.$$

Solution. Here $\Delta = 25+24 = 49 > 0$. Thus the denominator factorizes.

$$3x^2-5x-2 = (3x+1)(x-2).$$

Hence the function is not integrable for $x = -1/3$ or $x = 2$. Suppose

$$\frac{11x-1}{3x^2-5x-2} \equiv \frac{A}{3x+1} + \frac{B}{x-2}.$$

Then, $$11x-1 \equiv A(x-2) + B(3x+1).$$

Equating coefficients we obtain

$$A+3B = 11 \quad \text{and} \quad -2A+B = -1,$$

hence, $$A = 2 \quad \text{and} \quad B = 3.$$

Therefore,

$$I = \int \left\{\frac{2}{3x+1}+\frac{3}{x-2}\right\} dx,$$

$$= \frac{2}{3}\log|3x+1|+3\log|x-2|+C.$$

(2) $\Delta = 0$. The quadratic is therefore a perfect square and the function can be expressed in the form

$$\frac{A}{(px+q)^2}+\frac{B}{px+q}$$

which can easily be integrated (see 4.2 Ex. 1 and 2).

Examples

9. Integrate

$$I = \int \frac{dx}{9x^2-12x+4}.$$

Solution. $\Delta = 144-144 = 0$. The denominator is therefore a perfect square, and

$$9x^2-12x+4 = (3x-2)^2.$$

Hence, x cannot take the value 2/3.

$$I = \int \frac{dx}{(3x-2)^2} = \int (3x-2)^{-2}\, dx,$$

$$= -\frac{(3x-2)^{-1}}{3}+C,$$

$$= C-\frac{1}{3(3x-2)}.$$

10. Integrate

$$I = \int \frac{9x-5}{9x^2-6x+1}\, dx.$$

Solution. $\Delta = 36-36 = 0$. The denominator is therefore a perfect square and

$$9x^2-6x+1 = (3x-1)^2.$$

Hence, the function is not integrable for $x = \frac{1}{3}$. Suppose

$$\frac{9x-5}{(3x-1)^2} \equiv \frac{A}{(3x-1)^2} + \frac{B}{3x-1}.$$

Then $$9x-5 \equiv A+B(3x-1).$$

Equating coefficients we obtain

$$3B = 9 \quad \text{and} \quad A-B = -5,$$

hence, $$A = -2 \quad \text{and} \quad B = 3.$$

Therefore,

$$I = -2\int \frac{dx}{(3x-1)^2} + 3\int \frac{dx}{(3x-1)},$$

$$= -2\frac{(-1)}{3(3x-1)} + \log|3x-1| + C,$$

$$= \frac{2}{3(3x-1)} + \log|3x-1| + C.$$

(3) $\Delta < 0$. The quadratic expression cannot then be resolved into real factors so it is rearranged in the form $X^2 + A^2$ (where X is a linear function of x and A is independent of x) and integrated as an arc tan. If the numerator is a linear function of x it should be written as the sum of the derivative of the denominator (or a multiple of the derivative) and the appropriate compensating numerical term (see Ex. 12 below).

Examples

11. Integrate

$$I = \int \frac{dx}{2x^2-12x+27}.$$

Solution. $\Delta = 144\text{–}216 < 0$. Hence the quadratic has no real factors. We therefore rearrange the expression by completing the square and compensating the numerical term.

$$\begin{aligned}&2x^2-12x+27,\\&=2(x^2-6x)+27,\\&=2(x^2-6x+9)+27-18,\\&=2(x-3)^2+9.\end{aligned}$$

Therefore,

$$I=\frac{1}{2}\int\frac{dx}{(x-3)^2+9/2}.$$

Put $x-3 = X$, then $dx = dX$, and

$$\begin{aligned}I&=\frac{1}{2}\int\frac{dX}{X^2+9/2}\\&=\frac{1}{2}\cdot\frac{\sqrt{2}}{3}\text{ arc tan}\left(\frac{\sqrt{2}}{3}X\right)+C,\\&=\frac{1}{3\sqrt{2}}\text{ arc tan}\left(\frac{\sqrt{2}}{3}(x-3)\right)+C.\end{aligned}$$

12. Integrate

$$I=\int\frac{8x+5}{2x^2+6x+5}\,dx.$$

Solution. $\Delta = 36-40 < 0$. Hence the quadratic expression has no real factors and is integrable for all real x. We first express the numerator as the derivative of the denominator together with the appropriate numerical term.

$$\frac{d}{dx}(2x^2+6x+5) = 4x+6,$$

and

$$8x+5 = 2(4x+6)-7.$$

Hence,

$$\begin{aligned}I&=\int\frac{2(4x+6)-7}{2x^2+6x+5}\,dx,\\&=2\int\frac{4x+6}{2x^2+6x+5}\,dx-7\int\frac{dx}{2x^2+6x+5},\\&=2I_1-7I_2.\end{aligned}$$

$$I_1=\log(2x^2+6x+5).$$

Rewriting the denominator in I_2 by completing the square we obtain

$$\begin{aligned} &2x^2+6x+5, \\ &= 2(x^2+3x)+5, \\ &= 2\left(x+\frac{3}{2}\right)^2+\frac{1}{2}. \end{aligned}$$

Therefore,

$$\begin{aligned} I_2 &= \frac{1}{2}\int \frac{dx}{(x+3/2)^2+1/4} \\ &= \frac{1}{2}\cdot 2 \text{ arc tan } 2\left(x+\frac{3}{2}\right), \\ &= \text{arc tan } (2x+3). \end{aligned}$$

Hence, $$I = 2 \log (2x^2+6x+5)-7 \text{ arc tan } (2x+3)+C.$$

(iii) Alternatively any function of the type considered above may be integrated by expressing the denominator as the sum or differance of two squares. i.e. as $X^2 \pm A^2$, where X is a linear function of x and A (possibly zero) is independent of x. In effect this simply bypasses the necessity of examining the sign of the discriminant.

Example

13. Integrate

$$I = \int \frac{6\, dx}{x^2+2x-15}.$$

Solution. Rearranging the denominator we obtain

$$\begin{aligned} &x^2+2x-15 \\ &= (x+1)^2-1-15, \\ &= (x+1)^2-16. \end{aligned}$$

Hence,

$$I = \int \frac{6\, dx}{(x+1)^2-16}.$$

either $\int_{t_1}^{t_2} |h(t)|\, g'(t)\, \mathrm{d}t$ (for $x = g(t)$ monotonic increasing);

or $-\int_{t_1}^{t_2} |h(t)|\, g'(t)\, \mathrm{d}t$ (for $x = g(t)$ monotonic decreasing).

(b) If a curve is defined by the polar equation

$$r = f(\theta),$$

where $f(\theta)$ is a non-negative continuous function for $\alpha \leqslant \theta \leqslant \beta$, and $0 < (\beta - \alpha) < 2\pi$, then the area bounded by the arc $r = f(\theta)$ and the radial lines $\theta = \alpha$, $\theta = \beta$ (Fig. 73) is

$$A = \frac{1}{2}\int_{\alpha}^{\beta} r^2\, \mathrm{d}\theta.$$

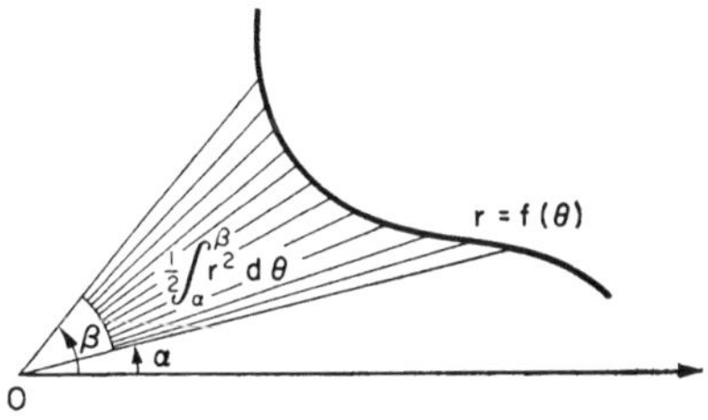

FIG. 73

Examples

1. Find the area of the ellipse (Fig. 74)

$$\left.\begin{aligned} x &= a\cos t \\ y &= b\sin t \end{aligned}\right\} \qquad \begin{aligned} &a > 0, \quad b > 0 \\ &0 \leqslant t \leqslant 2\pi. \end{aligned}$$

Solution. For $0 \leqslant t \leqslant \pi$, $a \cos t$ is monotonic decreasing, and for $\pi \leqslant t \leqslant 2\pi$, $a \cos t$ is monotonic increasing,

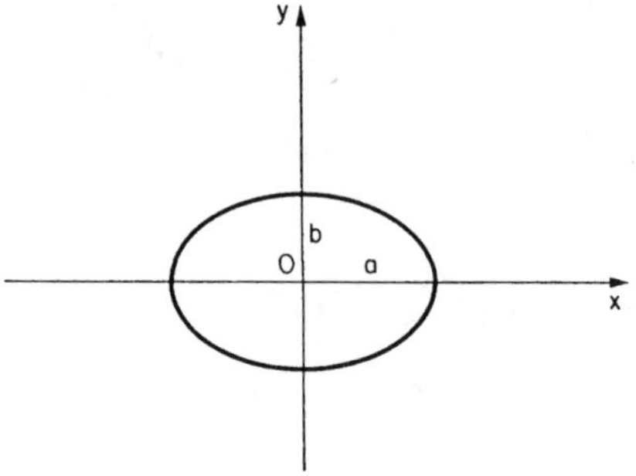

FIG. 74

therefore,
$$A = -\int_0^{\pi} |b \sin t| (-a \sin t)\, dt + \int_{\pi}^{2\pi} |b \sin t| (-a \sin t)\, dt.$$

In the first interval sin t is positive, and in the second interval sin t is negative, therefore

$$\begin{aligned} A &= ab\int_0^{\pi} \sin^2 t\, dt + ab\int_{\pi}^{2\pi} \sin^2 t\, dt, \\ &= ab\int_0^{2\pi} \sin^2 t\, dt, \\ &= ab\int_0^{2\pi} \frac{1}{2}(1-\cos 2t)\, dt, \\ &= \frac{1}{2}ab\left[t - \frac{1}{2}\sin 2t\right]_0^{2\pi}, \\ &= \pi ab. \end{aligned}$$

2. Find the area bounded by the x-axis and the cycloid

$$\left.\begin{aligned} x &= a(t-\sin t) \\ y &= a(1-\cos t) \end{aligned}\right\} \qquad a > 0$$

where $0 \leqslant t \leqslant 2\pi$ (Fig. 75).

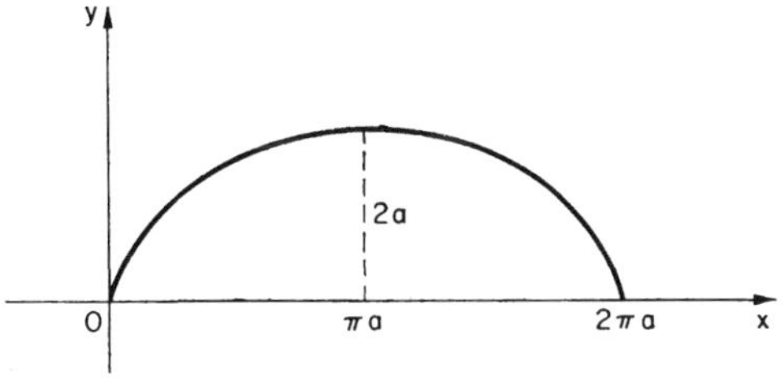

FIG. 75

Solution. Differentiating with respect to t we obtain

$$\dot{x} = a(1-\cos t).$$

Hence in the given interval

$$\dot{x} \geqslant 0,$$

and thus $x = a(t-\sin t)$ is monotonic increasing in the interval.

Therefore, $$A = \int_0^{2\pi} |a(1-\cos t)|\, a(1-\cos t)\, dt,$$

$$= a^2 \int_0^{2\pi} (1-\cos t)^2\, dt,$$

$$= a^2 \left[\frac{3}{2}t - 2\sin t + \frac{1}{4}\sin 2t\right]_0^{2\pi},$$

$$= 3\pi a^2.$$

3. Find the area of the astroid

$$\left.\begin{aligned} x &= a\cos^3 t \\ y &= a\sin^3 t \end{aligned}\right\} \qquad \begin{aligned} &a > 0 \\ &0 \leqslant t \leqslant 2\pi \text{ (Fig. 76).} \end{aligned}$$

Solution. For $0 \leqslant t \leqslant \pi$, $a\cos^3 t$ is decreasing, and for $\pi \leqslant t \leqslant 2\pi$, $a\cos^3 t$ is increasing, thus we can find the required area as in Example 1. But we notice that

(i) for $t = \pm t_0$, $x = x_0$ and $y = \pm y_0$, hence the curve is symmetrical about the x-axis;

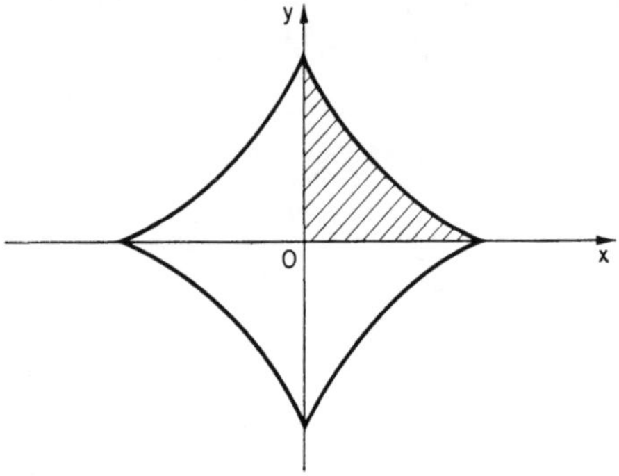

FIG. 76

(ii) for $t = t_0$ and $(\pi - t_0)$, $x = \pm x_0$, and $y = y_0$, hence the curve is symmetrical about the y-axis.

Therefore, the area in the first quadrant $(0 \leqslant t \leqslant \pi/2)$ is a quarter of the whole area.

Thus,
$$A = -4 \int_0^{\pi/2} |a \sin^3 t| \, (-3a \cos^2 t \sin t) \, dt,$$
$$= +12a^2 \int_0^{\pi/2} \sin^4 t \cos^2 t \, dt,$$
$$= 12a^2 \cdot \frac{3}{6} \cdot \frac{1}{4} \cdot \frac{1}{2} \cdot \frac{\pi}{2} \text{ (using the reduction formula)},$$
$$= \frac{3}{8} \pi a^2.$$

Note: We can eliminate t between the 2 parametric equations to obtain the cartesian equation of the astroid

$$x^{2/3} + y^{2/3} = a^{2/3}.$$

4. Find the area of the logarithmic spiral

$$r = ae^{k\theta}, \qquad a > 0, \qquad k > 0,$$

between the radial lines $\theta=0$, $\theta=\alpha$, where $0<\alpha\leqslant 2\pi$ (Fig. 77).

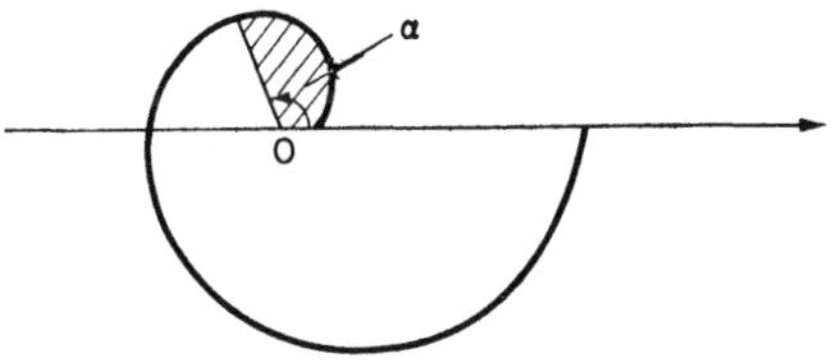

FIG. 77

Solution

$$A=\frac{1}{2}\int_0^\alpha r^2\,d\theta,$$

$$=\frac{1}{2}\int_0^\alpha a^2e^{2k\theta}\,d\theta,$$

$$=\frac{1}{4k}a^2\Big[e^{2k\theta}\Big]_0^\alpha.$$

Therefore, $$A=\frac{a^2}{4k}(e^{2k\alpha}-1).$$

5. Find the area of the lemniscate

$$r^2=a^2\cos 2\theta,\qquad a>0 \text{ (Fig. 78)}.$$

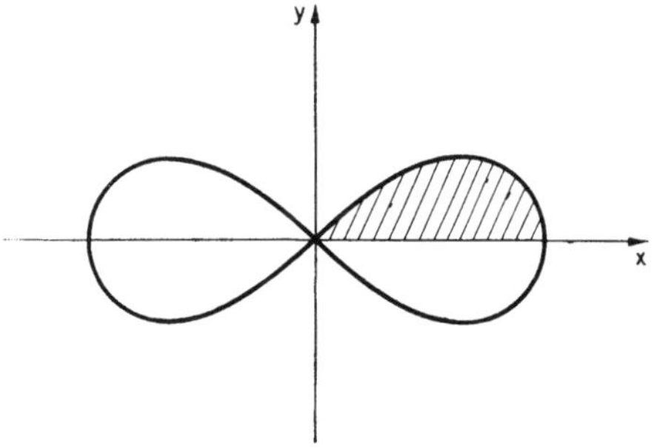

FIG. 78

Solution. Notice that for real $r_9 - \pi/4 \leqslant \theta \leqslant \pi/4$, and $3\pi/4 \leqslant \theta \leqslant 5\pi/4$. As in Example 3 above, we can show that the curve is symmetrical about the lines $\theta = 0$ and $\theta = \pi/2$. Thus by integrating through the interval $0 \leqslant \theta \leqslant \pi/4$ we obtain as before a quarter of the total area. In this way we find the area of the shaded portion of the figure, moving along the curve in the direction shown.

Therefore,
$$A = 4 \cdot \frac{1}{2} \int_0^{\pi/4} a^2 \cos 2\theta \, d\theta,$$
$$= 2a^2 \left[\frac{1}{2} \sin 2\theta \right]_0^{\pi/4},$$
$$= a^2.$$

6. Find the area enclosed by the curve

$$r = a \sin 3\theta, \qquad a > 0,$$

where $0 \leqslant \theta \leqslant \pi/3$, $2\pi/3 \leqslant \theta \leqslant \pi$, $4\pi/3 \leqslant \theta \leqslant 5\pi/3$. (Fig. 79).

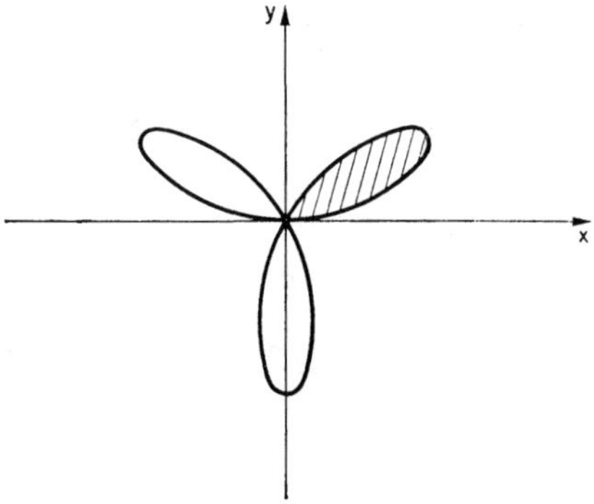

FIG. 79

Solution. In each of the given intervals $\sin 3\theta \geqslant 0$, thus $r \geqslant 0$. The curve has three lines of symmetry,

$$\theta = \frac{1}{6}\pi, \qquad \theta = \frac{5}{6}\pi, \qquad \theta = \frac{3}{2}\pi,$$

$$\left.\begin{aligned} &\text{since for} && \theta = \frac{1}{6}\pi \pm \theta_0, \\ &\text{or} && \theta = \frac{5}{6}\pi \pm \theta_0, \\ &\text{or} && \theta = \frac{3}{2}\pi \pm \theta_0 \end{aligned}\right\} \quad r = a\cos 3\theta_0.$$

Hence,

$$\begin{aligned} A &= 3 \cdot \frac{1}{2}\int_0^{\pi/3} r^2\, d\theta, \\ &= \frac{3}{2}a^2 \int_0^{\pi/3} \sin^2 3\theta\, d\theta, \\ &= \frac{3}{4}a^2 \left[\theta - \frac{1}{6}\sin 6\theta\right]_0^{\pi/3}. \end{aligned}$$

Therefore,

$$A = \frac{1}{4}\pi a^2.$$

EXERCISE 26

1. Find the area enclosed by the curve $r = a \sin \theta$ and the radial lines $\theta = 0$ and $\theta = \pi$. Sketch the curve.

2. Find the area enclosed by the curve $r = a \sin 2\theta$ and the radial lines $\theta = 0$, $\theta = 2\pi$.

3. Find the area enclosed by the curve $r = a \sin 6\theta$, where $a > 0$, and $0 \leqslant \theta \leqslant 2\pi$.

4. Find the area enclosed by the x-axis and the cardioid $r = a(1 + \cos \theta)$, where $a > 0$ and $0 \leqslant \theta \leqslant 2\pi$.

5. Find the area enclosed by the spiral of Archimedes $r = a\theta$, where $a > 0$, and $0 \leqslant \theta \leqslant 2\pi$.

6. Find the area of the circle $r = a \cos \theta$, $a > 0$, $0 \leqslant \theta \leqslant 2\pi$.

7. Find the area of the ellipse $\dfrac{p}{r} = 1 + e \cos \theta$.

8. Find the area of a loop of the curve $x = 3t^2$, $y = 3t - t^3$, where the parameter t lies in the interval $-\sqrt{3} \leqslant t \leqslant \sqrt{3}$.

9. Find the area of a loop of the curve

$$x = \frac{3at}{1+t^3}, \qquad y = \frac{3at^2}{1+t^3},$$

where $a > 0$, and the parameter t lies in the interval $0 \leqslant t \leqslant \infty$.

10. Find the area between the x-axis and the curve $x = 2a \sin^2 \Phi$, $y = 2a \sin^2 \Phi \tan \Phi$, for $\pi/12 \leqslant \Phi \leqslant \pi/6$.

11. Find the area of a loop of the curve $x = t^2$, $y = t + t^3/3$, $0 \leqslant t \leqslant \sqrt{3}$.

12. Find the area of a loop of the curve $x = 2t - t^2$, $y = 2t^2 - t^3$.

13. Find the area of the curve (the evolute of a circle) $x = 3(\cos t + t \sin t)$, $y = 3(\sin t - t \cos t)$, $0 \leqslant t \leqslant 2\pi$.

14. Find the area of a loop of the curve $x = 2(2 \cos t - \cos 2t)$, $y = 2(2 \sin t - \sin 2t)$.

3.2 The calculation of lengths of arc

(a) If the function $y = f(x)$, defined in the interval $a \leqslant x \leqslant b$, has a continuous first derivative in the interval, then the length of arc of the curve in this interval is

$$L = \int_a^b \sqrt{\left\{1 + \left(\frac{dy}{dx}\right)^2\right\}}\, dx,$$

and the element of arc is

$$dL = \sqrt{\left\{1 + \left(\frac{dy}{dx}\right)^2\right.}\, dx.$$

(b) If a curve is defined parametrically by the equations $x = g(t)$, $y = h(t)$, and in the interval $t_1 \leqslant t \leqslant t_2$, the functions $g(t)$ and $h(t)$ have continuous first derivatives and are single-valued, then the length of arc in the interval is

$$L = \int_{t_1}^{t_2} \sqrt{(\dot{x}^2 + \dot{y}^2)}\, dt,$$

and

$$dL = \sqrt{(\dot{x}^2 + \dot{y}^2)}\, dt.$$

(c) If a curve is defined by its polar equation $r = f(\theta)$, and in the interval $\alpha \leqslant \theta \leqslant \beta$ the function $f(\theta)$ has a continuous first derivative and is single-valued, then

$$L = \int_\alpha^\beta \sqrt{\left\{r^2 + \left(\frac{dr}{d\theta}\right)^2\right\}}\, d\theta,$$

and

$$dL = \sqrt{\left\{r^2+\left(\frac{dr}{d\theta}\right)^2\right\}}\, d\theta.$$

[Note that if $\beta-\alpha \leqslant \pi$ (or if $\beta-\alpha \leqslant 2\pi$ and $r \geqslant 0$), then $r=f(\theta)$ is certainly single-valued in the interval.]

Examples

7. Find the length of arc of the parabola $y=x^2$ in the interval $0 \leqslant x \leqslant 2$.

Solution

$$\frac{dy}{dx} = 2x,$$

therefore, $$L = \int_0^2 \sqrt{(1+4x^2)}\, dx,$$

$$= \left[\frac{1}{2}x\sqrt{(1+4x^2)}+\frac{1}{4}\log\{2x+\sqrt{(1+4x^2)}\}\right]_0^2,$$

$$= \sqrt{17}+\frac{1}{4}\log(4+\sqrt{17}).$$

(See Chapter 10, 5.4 for evaluation of the integral.)

8. Find the length of arc of the astroid

$$\left.\begin{aligned} x &= a\cos^3 t \\ y &= a\sin^3 t \end{aligned}\right\} \qquad a>0, \quad 0 \leqslant t \leqslant 2\pi.$$

Solution

$$\dot{x} = -3a\cos^2 t\sin t,$$

$$\dot{y} = 3a\sin^2 t\cos t.$$

Thus, $$L = \int_0^{2\pi} \sqrt{(9a^2\cos^4 t\sin^2 t+9a^2\sin^4 t\cos^2 t)}\, dt,$$

$$= 3a\int_0^{2\pi} |\sin t\cos t|\sqrt{(\cos^2 t+\sin^2 t)}\, dt,$$

$$= \frac{3a}{2}\int_0^{2\pi} |\sin 2t|\, dt.$$

We have already shown that the astroid is symmetrical about the x- and y-axes, hence

$$L = 4\cdot\frac{3a}{2}\int_0^{\pi/2} \sin 2t\, dt,$$
$$= 6a\left[-\frac{1}{2}\cos 2t\right]_0^{\pi/2},$$
$$= 6a.$$

9. Find the length of the arc of the cycloid

$$\left.\begin{aligned} x &= a(t-\sin t)\\ y &= a(1-\cos t)\end{aligned}\right\} \qquad a > 0,$$

for $0 \leqslant t \leqslant 2\pi$.

Solution

$$\dot{x} = a(1-\cos t),$$
$$\dot{y} = a \sin t.$$

Hence, $$L = \int_0^{2\pi} \sqrt{\{a^2(1-\cos t)^2 + a^2 \sin^2 t\}}\, dt,$$
$$= a\int_0^{2\pi} \sqrt{(1-2\cos t+\cos^2 t+\sin^2 t)}\, dt,$$
$$= a\int_0^{2\pi} \sqrt{\{2(1-\cos t)\}}\, dt,$$
$$= a\int_0^{2\pi} \sqrt{\left\{2\left(2\sin^2 \frac{1}{2}t\right)\right\}}\, dt,$$
$$= 2a\int_0^{2\pi} \left|\sin \frac{1}{2}t\right| dt.$$

In this interval $\sin \frac{1}{2} t$ is positive, hence

$$L = 2a\int_0^{2\pi} \sin \frac{1}{2} t\, dt,$$
$$= 2a\left[-2\cos \frac{1}{2}t\right]_0^{2\pi}.$$

Therefore, $$L = 8a.$$

10. Find the length of the logarithmic spiral

$$r = ae^{k\theta}, \qquad a > 0, \qquad k \neq 0, \qquad 0 \leqslant \theta \leqslant \alpha.$$

Solution

$$\frac{dr}{d\theta} = ake^{k\theta}.$$

Hence,

$$L = \int_0^\alpha \sqrt{(a^2e^{2k\theta} + a^2k^2e^{2k\theta})}\, d\theta,$$

$$= a\sqrt{(1+k^2)} \int_0^\alpha e^{k\theta}\, d\theta,$$

$$= \frac{a}{k}\sqrt{(1+k^2)} \Big[e^{k\theta} \Big]_0^\alpha,$$

$$= \frac{a}{k}\sqrt{(1+k^2)}\,(e^{k\alpha} - 1).$$

EXERCISE 27

Find the lengths of arc of the following curves:

1. An arc of the semi-cubical parabola $y^2 = 4x^3$, $y > 0$.
2. $9y^2 = 4x^3, \quad 0 \leqslant x \leqslant 3.$
3. $9y^2 = 2x^3, \quad 0 \leqslant x \leqslant 2.$
4. $3y^2 = 4x^3, \quad 0 \leqslant x \leqslant 1.$
5. $9y^2 = x^3, \quad 0 \leqslant x \leqslant 12.$
6. $2y^2 = 3x^3, \quad 0 \leqslant x \leqslant 2.$
7. $2y^2 = x^3, \quad 0 \leqslant x \leqslant 2.$
8. $y = 2\sqrt{x}, \quad A(1, 2), B(9, 6).$
9. $y^2 = 2x - x^2, \quad 0 \leqslant x \leqslant 1.$
10. $2y^2 = x - 2x^2, \quad 0 \leqslant x \leqslant 1/2.$
11. $y = 2\sqrt{x}, \quad 0 \leqslant x \leqslant 1.$
12. $y = 2\sqrt{3x}, \quad 0 \leqslant x \leqslant 1.$
13. $y = \frac{2}{3}(x-1)^{3/2}, \quad 1 \leqslant x \leqslant 4.$
14. $y = \log \sin x, \quad \frac{1}{3}\pi \leqslant x \leqslant \frac{1}{2}\pi.$
15. $y = 1 - \log \cos x, \quad 0 \leqslant x \leqslant \frac{1}{4}\pi.$
16. $y = c \cosh \frac{x}{c}, \quad -c \leqslant x \leqslant c.$
17. $y = \log x, \quad \sqrt{3} \leqslant x \leqslant 2\sqrt{2}.$

18. $y = \log(1-x^2), \quad 0 \leqslant x \leqslant \frac{1}{2}.$

19. $y = -\log \tanh \frac{x}{2}, \quad 1 \leqslant x \leqslant 2.$

20. $y = \arcsin x + \sqrt{(1-x^2)}, \quad -1 \leqslant x \leqslant 1.$

21. $y^2 = \frac{x^3}{2a-x}, \quad 0 \leqslant x \leqslant \frac{5}{3} a.$

22. $x = \frac{1}{4} y^2 - \frac{1}{2} \log y, \quad 1 \leqslant y \leqslant 2.$

23. $y = \mathrm{e}^x, \quad A(0, 1), \quad B(1, \mathrm{e}).$

24. $r = a(1+\cos\theta), \quad 0 \leqslant \theta \leqslant \pi.$

25. $r = a\theta, \quad 0 \leqslant \theta \leqslant 1.$

26. $r\theta = a, \quad \frac{2}{3} \leqslant \theta \leqslant \frac{3}{4}.$

27. $r = 2a \cos\theta.$

28. $\frac{p}{r} = 1 + \cos\theta, \quad -\frac{\pi}{2} \leqslant \theta \leqslant \frac{\pi}{2}.$

29. $x = t^2, \quad y = t - \frac{1}{3} t^3, \quad 0 \leqslant t \leqslant \sqrt{3}.$

30. $x = a(\cos t + t \sin t), \quad y = a(\sin t - t\cos t), \quad 0 \leqslant t \leqslant \pi.$

31. $x = a \cos^5 \frac{t}{2}, \quad y = a \sin^5 \frac{t}{2}, \quad 0 \leqslant t \leqslant \pi.$

32. $x = 2a \sin^2 t, \quad y = 2a \sin^2 t \tan t, \quad 0 \leqslant t \leqslant \frac{\pi}{3}.$

33. $x = 5 \cos t(1+\cos t), \quad y = 5 \sin t(1+\cos t), \quad 0 \leqslant t \leqslant 2\pi.$

34. $x = a \cos^4 t, \quad y = a \sin^4 t, \quad 0 \leqslant t \leqslant \frac{1}{2}\pi.$

35. Given that in three dimensions the length of a curve is

$$L = \int_{t_1}^{t_2} \sqrt{(\dot{x}^2 + \dot{y}^2 + \dot{z}^2)}\, \mathrm{d}t,$$

find the length of the curve $x = 3\cos t$, $y = 3 \sin t$, $z = 4t$, $0 \leqslant t \leqslant 2\pi$.

3.3 Volumes of revolution and surface area

(a) If the function $y = f(x)$ is continuous and non-negative in the closed interval $[a, b]$, then the volume of the solid generated by rotating the arc AB and the ordinates $x = a$ and $x = b$

about the x-axis (Fig. 80) is

$$V = \pi \int_a^b y^2 \, dx.$$

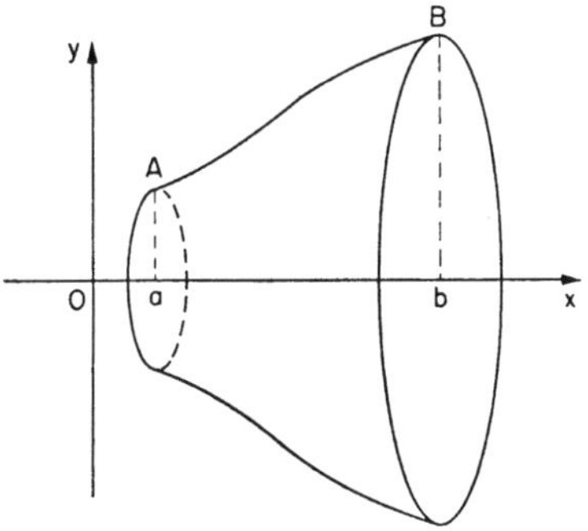

FIG. 80

(b) The area of the curved surface of the solid so formed is

$$S = 2\pi \int_a^b y \, dL,$$

$$= 2\pi \int_a^b y \sqrt{\left\{1+\left(\frac{dy}{dx}\right)^2\right\}} \, dx,$$

assuming that dy/dx is continuous in the given interval.

(c) Similarly, under the conditions stated in the previous paragraph, if $x = g(t)$, $y = h(t)$, $t_1 \leqslant t \leqslant t_2$,

$$V = \pi \int_{t_1}^{t_2} y^2 \dot{x} \, dt,$$

and $$S = 2\pi \int_{t_1}^{t_2} y \sqrt{(\dot{x}^2+\dot{y}^2)} \, dt.$$

Examples

11. Find the surface area and the volume of the solid formed by revolving the parabola $y^2 = 4x$, between $x = 0$ and $x = 3$, about the x-axis.

Solution. We first find the element of arc length dL.

$$\frac{dy}{dx} = \frac{1}{\sqrt{x}},$$

therefore,

$$dL = \sqrt{\left(1+\frac{1}{x}\right)}\, dx.$$

Hence the surface area is

$$S = 2\pi \int_0^3 2\sqrt{x}\sqrt{\left(1+\frac{1}{x}\right)}\, dx,$$

$$= 4\pi \int_0^3 \sqrt{(x+1)}\, dx,$$

$$= 4\pi \cdot \frac{2}{3}\Big[(x+1)^{3/2}\Big]_0^3,$$

$$= \frac{8}{3}\pi(4^{3/2}-1^{3/2}),$$

$$= \frac{56}{3}\pi.$$

And the volume of the solid is

$$V = \pi \int_0^3 y^2\, dx,$$

$$= \pi \int_0^3 4x\, dx,$$

$$= 2\pi\Big[x^2\Big]_0^3,$$

$$= 18\pi.$$

12. Find the surface area and volume of the solid obtained by revolving the cycloid

$$x = a(t-\sin t), \qquad y = a(1-\cos t),$$

for $0 < t < 2\pi$, about the x-axis.

Solution. We notice that $\dot{x} = a(1-\cos t)$ is not negative, and that consequently x is an increasing function of t. Also y is non-negative for all t.

The volume V is given by,

$$V = \pi \int_0^{2\pi} a^2(1-\cos t)^2 \cdot a(1-\cos t)\,\mathrm{d}t,$$

$$= \pi a^3 \int_0^{2\pi} (1-3\cos t+3\cos^2 t-\cos^3 t)\,\mathrm{d}t,$$

$$= \pi a^3 \left[t-3\sin t+3\left(\frac{1}{2}t+\frac{1}{4}\sin 2t\right) - \left(\sin t-\frac{1}{3}\sin^3 t\right)\right]_0^{2\pi},$$

$$= 5\pi^2 a^3.$$

The surface area is,

$$S = 2\pi \int_0^{2\pi} a(1-\cos t)\sqrt{\{a^2(1-\cos t)^2+a^2\sin^2 t\}}\,\mathrm{d}t,$$

$$= 2\pi \int_0^{2\pi} a(1-\cos t)\cdot a\sqrt{\{2(1-\cos t)\}}\,\mathrm{d}t,$$

$$= 2\pi \int_0^{2\pi} a(1-\cos t)\cdot 2a\sin\frac{1}{2}t\,\mathrm{d}t,$$

$$= 8\pi a^2 \int_0^{2\pi} \sin^3\frac{1}{2}t\,\mathrm{d}t,$$

$$= 16\pi a^2 \int_0^{\pi} \sin^3 x\,\mathrm{d}x, \qquad \left(\text{where } x = \frac{1}{2}t\right)$$

$$= 32\pi a^2 \int_0^{\pi/2} \sin^3 x\,\mathrm{d}x,$$

$$= 32\,\pi a^2\cdot\frac{2}{3}\cdot 1,$$

$$= \frac{64}{3}\pi a^2.$$

13. Find the surface area and the volume of the solid generated by revolving the ellipse $x^2/a^2+y^2/b^2=1$ about the x-axis.

Solution

$$V=\pi\int_{-a}^{a} y^2\,\mathrm{d}x,$$

then, $$\frac{1}{2}V=\pi\int_{0}^{a}\left(b^2-\frac{b^2}{a^2}x^2\right)\mathrm{d}x,$$

and $$V=2\pi b^2\left[x-\frac{1}{a^2}\cdot\frac{x^3}{3}\right]_0^a,$$

$$=\frac{4}{3}\pi ab^2.$$

In order to find the surface area it will be simpler to express the ellipse in its parametric form,

$$x=a\cos t,\qquad y=b\sin t,\qquad 0\leqslant t\leqslant\pi.$$

In the given interval x is a decreasing function and y is non-negative. Then,

$$S=2\pi\int_{0}^{\pi} b\sin t\,\sqrt{(a^2\sin^2 t+b^2\cos^2 t)}\,\mathrm{d}t.$$

Put $\cos t=u$, then $-\sin t\,\mathrm{d}t=\mathrm{d}u$, and the interval of integration is $(-1, 1)$, then

$$S=2\pi b\int_{-1}^{1}\sqrt{\{a^2(1-u^2)+b^2u^2\}}\,\mathrm{d}u,$$

$$=2\pi b\int_{-1}^{1}\sqrt{\{a^2-(a^2-b^2)u^2\}}\,\mathrm{d}u.$$

(i) If $a=b$,

$$S=4\pi a^2.$$

(ii) Otherwise put $\dfrac{a^2-b^2}{a^2}=\varepsilon^2.$

Then,

$$S = 2\pi ab \int_{-1}^{1} \sqrt{(1-\varepsilon^2 u^2)}\, du,$$

$$= 2\pi ab \left[\frac{1}{2} u\sqrt{(1-\varepsilon^2u^2)} + \frac{1}{2\varepsilon} \text{ arc sin } \varepsilon u \right]_{-1}^{1},$$

$$= 2\pi ab \left\{ \sqrt{(1-\varepsilon^2)} + \frac{1}{\varepsilon} \text{ arc sin } \varepsilon \right\}.$$

If $a > b$, then

$$S = 2\pi ab \left\{ \frac{b}{a} + \frac{a}{\sqrt{(a^2-b^2)}} \text{ arc sin } \frac{\sqrt{(a^2-b^2)}}{a} \right\}.$$

If $a < b$, then $\varepsilon = i\dfrac{\sqrt{(b^2-a^2)}}{a} = i\alpha.$

Thus,
$$\frac{1}{\varepsilon} \text{ arc sin } \varepsilon = \frac{1}{i\alpha} \text{ arc sin } i\alpha,$$

$$= \frac{1}{\alpha} \sinh^{-1}\alpha.$$

Therefore, $S = 2\pi ab \left\{ \dfrac{b}{a} + \dfrac{a}{\sqrt{(b^2-a^2)}} \sinh^{-1} \dfrac{\sqrt{(b^2-a^2)}}{a} \right\},$

or in logarithmic form

$$S = 2\pi ab \left\{ \frac{b}{a} + \frac{a}{\sqrt{(b^2-a^2)}} \log \frac{b+\sqrt{(b^2-a^2)}}{a} \right\}.$$

EXERCISE 28

1. Find the surface area and volume of revolution of the solid obtained by revolving the curve $xy^2 = 1$ about the x-axis between $x = a$ and $x = b$ (where $b > a > 0$).

2. Find the volume of the solid obtained by revolving the hyperbola $xy=1$ about the x-axis, for $1 \leqslant x \leqslant \infty$.

3. Find the surface area and volume of the solid obtained by revolving the sine curve $y = \sin x$ about the x-axis for $0 \leqslant x \leqslant \pi$.

4. Find the volume of the solid obtained by revolving the curve $y^2 (x-4) = x(x-3)$ about the x-axis, $0 \leqslant x \leqslant 3$.

5. Find the surface area and volume of the solid generated by rotating the hyperbola $x^2-y^2 = a^2$ about the x-axis, for $a \leqslant x \leqslant a\sqrt{2}$.

6. Find the surface area and volume of the solid generated by rotating the parabola $3y-x^2 = 0$, $0 \leqslant x \leqslant 1$, about the x-axis.

7. Find the surface area and volume of the solid obtained by rotating the catenary $y = c \cosh \dfrac{x}{c}$ about the x-axis, for $-c \leqslant x \leqslant c$.

8. Find the surface area of the solid obtained by revolving the curve $y = \sqrt{(2ax-x^2)}$ about the x-axis, $(0 \leqslant x \leqslant 2)$.

9. Find the surface area and volume of the solid obtained by rotating the cissoid $x^3 = y^2(2a-x)$ about the x-axis where $1 \leqslant x \leqslant 2$.

Find the volumes of the solids generated by rotating the following curves about the x-axis in the given ranges:

10. $y = a \cos \dfrac{2\pi}{b} x, \quad 0 \leqslant x \leqslant b/4.$

11. $y = \sin^{7/2} x, \quad 0 \leqslant x \leqslant \pi/2.$

12. $y = \dfrac{\sqrt{(3x+1)}}{\sqrt[4]{(x^2-6x+15)}}, \quad 0 \leqslant x \leqslant 1.$

13. The ellipse $4x^2+9y^2 = 36$.

14. $3y^2 = 4x, \quad 0 \leqslant x \leqslant 1.$

15. The ellipse $16x^2+9y^2 = 144$.

16. $y = 2x^3, \quad 0 \leqslant x \leqslant 1.$

17. $y = \cos^{7/2} x, \quad 0 \leqslant x \leqslant \pi/2.$

18. The hyperbola $y(x-1) = 1, \quad 2 \leqslant x \leqslant 4.$

19. $y = \dfrac{1}{\sqrt{(x^2-1)}}, \quad 2 \leqslant x \leqslant 4.$

20. $25x^2+4y^2 = 100.$

21. The circle $x^2+y^2-20y+75 = 0.$

22. $y = e^{-x} \sqrt{(\sin x)}, \quad 0 \leqslant x \leqslant 2\pi.$

23. Find the volume and surface area of the solid generated by rotating the astroid $x = a \cos^3 t$, $y = a \sin^3 t$, $0 \leqslant t \leqslant \pi$, about the x-axis.

24. Find the volume and surface area of the solid obtained by revolving the curve $x = t^2$, $y = t-t^3/3$, $0 \leqslant t \leqslant \sqrt{3}$, about the x axis.

3.4 Moment of inertia and centre of gravity

(a) *Definitions*

(i) If masses $m_1, m_2, \ldots, m_k$ are at points (x_1, y_1), (x_2, y_2), $\ldots$, (x_k, y_k), then the moment of inertia of the masses about the x-axis is

$$I_x = \sum_{i=1}^{k} m_i y_i^2,$$

and the moment of inertia of the system about the y-axis is

$$I_y = \sum_{i=1}^{k} m_i x_i^2.$$

(ii) The moment of inertia of an arc AB of a curve $y = f(x)$ (Fig. 81) about the x-axis is

$$I_x = \int_{(A)}^{(B)} my^2 \, dL,$$

where m is the mass per unit length of the line.

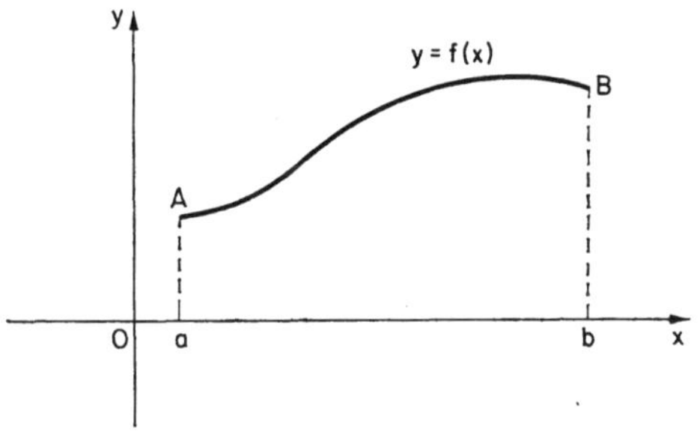

FIG. 81

(iii) If $y = f(x)$ is continuous for $a \leqslant x \leqslant b$, then the moment of inertia about the x-axis of the area bounded by $y = f(x)$ and the lines $x = a$ and $x = b$ is

$$I_x = \frac{1}{3} \int_a^b my^3 \, dx,$$

where m is the mass per unit area.

(iv) The moment of inertia about the x-axis of the solid generated by revolving the curve $y = f(x)$ about the x-axis between $x = a$ and $x = b$ (Fig. 82) is

$$I_x = \frac{1}{2}\pi \int_a^b my^4 \, dx,$$

where m is the mass per unit volume.

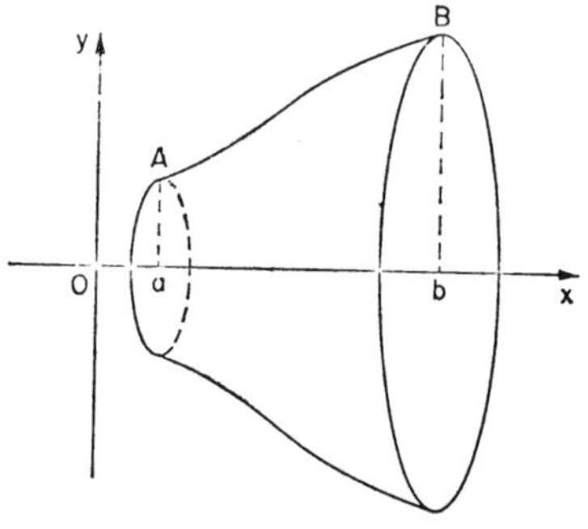

FIG. 82

(v) If masses $m_1, m_2, \ldots, m_k$ are at points (x_1, y_1), (x_2, y_2), $\ldots$, (x_k, y_k), then the coordinates $(\bar{x}, \bar{y})$ of the centre of gravity of the system are

$$\bar{x} = \frac{\sum_{i=1}^{k} m_i x_i}{\sum_{i=1}^{k} m_i}, \qquad \bar{y} = \frac{\sum_{i=1}^{k} m_i y_i}{\sum_{i=1}^{k} m_i}.$$

(vi) The coordinates of the centre of gravity of an arc of a curve of uniform mass per unit length (Fig. 81) are

$$\bar{x} = \frac{\int_{(A)}^{(B)} x \, dL}{\int_{(A)}^{(B)} dL}, \qquad \bar{y} = \frac{\int_{(A)}^{(B)} y \, dL}{\int_{(A)}^{(B)} dL}.$$

(vii) The coordinates of the centre of gravity of the area bounded by $y = f(x)$, $x = a$ and $x = b$, of uniform mass per unit area (Fig. 81) are

$$\bar{x} = \frac{\int_a^b xy\,dx}{\int_a^b y\,dx}, \qquad \bar{y} = \frac{\int_a^b y^2/2\,dx}{\int_a^b y\,dx}.$$

(viii) If $y = f(x)$ revolves about the x-axis between $x = a$ and $x = b$, the coordinates of the centre of gravity of the solid of uniform mass per unit volume so obtained are

$$\bar{x} = \frac{\int_a^b xy^2\,dx}{\int_a^b y^2\,dx}, \qquad \bar{y} = 0.$$

(b) *Routh's rule for the calculation of moments of inertia.* The moment of inertia of a certain number of uniform bodies of regular shape may be found according to the following rule (due to E. J. Routh):

The moment of inertia of the body about an axis of symmetry of the body

$$= \text{mass} \times \frac{\text{sum of squares of semi-axes perpendicular to this axis}}{3,\ 4,\ \text{or}\ 5},$$

according as the semi-axes terminate in a plane, elliptic, or ellipsoidal surface.

For example, for a circular solid cylinder of length $2l$ and radius r, the moment of inertia about the central axis is

$$M\left(\frac{r^2+r^2}{4}\right) = \frac{1}{2}Mr^2,$$

where M is the mass of the cylinder. About an axis through the centre perpendicular to the central axis, the moment of inertia is

$$M\left(\frac{r^2}{4}+\frac{l^2}{3}\right).$$

It should be noted, however, that the rule is applicable only to certain simple regular bodies.

(c) *The theorems of Pappus.* These are two theorems which concern the surface area and volume of solids of revolution.

(1) If an arc of a plane curve $y = f(x)$ revolves about an axis in its plane which does not cross the arc, the area of the surface generated is equal to the length of the arc multiplied by the length of the path traced out by the centre of mass of the arc. Thus if the length of the arc is L and the coordinates of the centre of gravity are $(\bar{x}, \bar{y})$, then

$$S = 2\pi\bar{y}L.$$

(2) If a plane closed curve revolves about an axis in its plane which does not intersect the curve, the volume of the solid generated is equal to the area enclosed by the curve multiplied by the length of the path traced out by the centre of gravity of the area. Thus if the area enclosed by the curve $y = f(x)$ is A, and the coordinates of the centre of gravity of the area are $(\bar{x}, \bar{y})$, then

$$V = 2\pi\bar{y}A,$$

where V is the volume generated by revolving the curve about the x-axis.

Examples

15. Find the moment of inertia, about the x-axis of the arc of the astroid

$$x^{2/3} + y^{2/3} = a^{2/3}, \qquad 0 \leqslant x \leqslant a,$$

which is above the x-axis, assuming that the mass per unit length is uniform.

Solution

$$y = (a^{2/3} - x^{2/3})^{3/2},$$

therefore,

$$y' = \frac{3}{2}(a^{2/3} - x^{2/3})^{1/2}\left(-\frac{2}{3}x^{-1/3}\right)$$

$$= -\frac{(a^{2/3} - x^{2/3})^{1/2}}{x^{1/3}}.$$

Then, $$\mathrm{d}L = \sqrt{\left\{1+\frac{a^{2/3}-x^{2/3}}{x^{2/3}}\right\}}\,\mathrm{d}x$$
$$= \left(\frac{a}{x}\right)^{1/3}\mathrm{d}x.$$

Therefore, $I_x = \int_{(A)}^{(B)} my^2\,\mathrm{d}L$, where m is mass per unit length,

$$= m\int_0^a \left(a^{2/3}-x^{2/3}\right)^3 a^{1/3}x^{-1/3}\,\mathrm{d}x,$$

$$= ma^{1/3}\left[\frac{3}{2}a^2x^{2/3}-\frac{9}{4}a^{4/3}x^{4/3} + \frac{3}{2}a^{2/3}x^2-\frac{3}{8}x^{8/3}\right]_0^a$$

$$= \frac{3}{8}ma^3.$$

But if M is the mass of the given length of arc,

$$M = m\int_{(A)}^{(B)}\mathrm{d}L,$$

$$= ma^{1/3}\int_0^a x^{-1/3}\,\mathrm{d}x,$$

$$= ma^{1/3}\left[\frac{3}{2}x^{2/3}\right]_0^a,$$

$$= \frac{3}{2}ma.$$

Therefore, $$I_x = \frac{1}{4}Ma^2.$$

16. Find the moment of inertia about the x-axis of the arc of the cycloid

$$x = a(t-\sin t), \qquad y = a(1-\cos t),$$

defined by the range $0 \leqslant t \leqslant 2\pi$.

Solution. For the cycloid

$$\mathrm{d}L = 2a \sin \frac{1}{2} t \, \mathrm{d}t.$$

Thus,

$$I_x = m \int_0^{2\pi} a^2(1 - \cos t)^2 \, 2a \sin \frac{1}{2} t \, \mathrm{d}t,$$

$$= 8ma^3 \int_0^{2\pi} \sin^5 \frac{1}{2} t \, \mathrm{d}t.$$

Put $\frac{1}{2} t = u$, then,

$$I_x = 16ma^3 \int_0^{\pi} \sin^5 u \, du,$$

$$= 32ma^3 \int_0^{\pi/2} \sin^5 u \, du,$$

$$= 32ma^3 \cdot \frac{4}{5} \cdot \frac{2}{3} \cdot 1,$$

$$= \frac{256}{15} ma^3.$$

If M is the mass of the given length of arc,

$$M = 2ma \int_0^{2\pi} \sin \frac{1}{2} t \, \mathrm{d}t,$$

$$= 8ma.$$

Therefore, $$I_x = \frac{32}{15} Ma^2.$$

17. Find the moment of inertia about the x-axis of the area bounded by the parabola $y^2 = 4ax$ and the ordinates $x = 1$ and $x = 4$.

Solution

$$I_x = \frac{1}{3} m \int_1^4 y^3 \, dx,$$ where m is the mass per unit area, thus

$$I_x = \frac{1}{3} m \int_1^4 (4ax)^{3/2} \, dx$$

$$= \frac{8}{3} ma^{3/2} \left[\frac{2}{5} x^{5/2} \right]_1^4$$

$$= \frac{496}{15} ma^{3/2}.$$

18. Find the moment of inertia about its axle of a wheel of internal radius r, external radius R, thickness a, and mass M.

Solution. Consider an elementary portion of the wheel, consisting of a hoop of radius x, width δx, and thickness a. If δM is the mass of the hoop, then the moment of inertia of the hoop about the axle is

$$\delta M x^2.$$

But, $\delta M = m . 2\pi x a \, \delta x$, where m is the mass per unit volume,

therefore, $$I = \int_r^R x^2 \, dM$$

$$= 2\pi ma \int_r^R x^3 \, dx$$

$$= \frac{1}{2} \pi ma(R^4 - r^4).$$

But, $$M = \int_r^R dM$$

$$= 2\pi ma \int_r^R x \, dx$$

$$= \pi ma \, (R^2 - r^2),$$

therefore, $$I = \frac{1}{2} M(R^2 + r^2).$$

19. Find the moment of inertia of the solid generated by revolving the parabola $y^2 = 4ax$ about the x-axis between $x = 0$ and $x = b$.

Solution

$$I_x = \frac{1}{2}\pi m \int_0^b y^4 \, dx,$$

$$= 8\pi m a^2 \int_0^b x^2 \, dx,$$

$$= \frac{8}{3}\pi m a^2 b^3.$$

If M is the mass of the given volume,

$$M = \pi m \int_0^b y^2 \, dx,$$

$$= 4\pi m a \int_0^b x \, dx,$$

$$= 2\pi m a b^2.$$

Therefore, $$I_x = \frac{4}{3} M a b.$$

20. Find the centre of gravity of the arc of the astroid $x^{2/3} + y^{2/3} = a^{2/3}$, which lies in the first quadrant.

Solution. The centre of gravity of an arc AB has coordinates,

$$\bar{x} = \frac{1}{L}\int_{(A)}^{(B)} x \, dL, \qquad \bar{y} = \frac{1}{L}\int_{(A)}^{(B)} y \, dL,$$

where $$L = \int_{(A)}^{(B)} dL.$$

We have already obtained the elementary arc length for an astroid (Ex. 15 above), namely

$$dL = \left(\frac{a}{x}\right)^{1/3} dx,$$

hence,
$$L = \int_0^a \left(\frac{a}{x}\right)^{1/3} dx,$$
$$= \frac{3}{2}a.$$

Therefore,
$$\bar{x} = \frac{2}{3a}\int_0^a x\left(\frac{a}{x}\right)^{1/3} dx,$$
$$= \frac{2}{3a}.a^{1/3}\left[\frac{3}{5}x^{5/3}\right]_0^a,$$
$$= \frac{2}{5}a.$$

From the symmetry of the equation of the astroid we may immediately deduce that the y coordinate of the centre of gravity is the same as the x coordinate.

Thus,
$$\bar{x} = \frac{2}{5}a, \qquad \bar{y} = \frac{2}{5}a.$$

21. Find the centre of gravity of that part of the ellipse $x^2/a^2+y^2/b^2 = 1$, which lies in the first quadrant.

Solution. We give two methods of solution.

(1)
$$\int_0^a y\,dx = \frac{b}{a}\int_0^a \sqrt{(a^2-x^2)}\,dx,$$
$$= \frac{b}{a}\int_0^{\pi/2} a^2\cos^2 t\,dt,$$
$$= \frac{1}{4}\pi ab.$$

$$\int_0^a xy\,dx = \frac{b}{a}\int_0^a x\sqrt{(a^2-x^2)}\,dx,$$
$$= \frac{b}{a}\left[\frac{1}{2}\left(-\frac{2}{3}\right)(a^2-x^2)^{3/2}\right]_0^a,$$
$$= \frac{1}{3}a^2b.$$

$$\frac{1}{2}\int_0^a y^2\,dx = \frac{b}{2a}\int_0^a (a^2-x^2)\,dx,$$

$$= \frac{b}{2a}\left[a^2x-\frac{1}{3}x^3\right]_0^a,$$

$$= \frac{1}{3}ab^2.$$

Then, quoting the formula for the coordinates of the centre of gravity,

$$\bar{x} = \frac{a^2b/3}{\pi ab/4}, \qquad \bar{y} = \frac{ab^2/3}{\pi ab/4},$$

that is,

$$\bar{x} = \frac{4a}{3\pi}, \qquad \bar{y} = \frac{4b}{3\pi}.$$

(2) Alternatively we may find the coordinates using Pappus' theorem.

The area of a quarter of the ellipse is $\pi ab/4$. The volume obtained by revolving this area about the x-axis is $2\pi ab^2/3$ (see Ex. 13 above). Hence, since $V = 2\pi\bar{y}A$,

$$\frac{2}{3}\pi ab^2 = 2\pi\bar{y}\cdot\frac{1}{4}\pi ab,$$

thus,

$$\bar{y} = \frac{4b}{3\pi}.$$

By symmetry we can immediately write

$$\bar{x} = \frac{4a}{3\pi}.$$

22. Find the centre of gravity of the solid obtained by revolving an arc of the parabola $y^2 = 4ax$ about the x-axis, where $0 \leqslant x \leqslant b$.

Solution. Clearly $\bar{y} = 0$.

$$\bar{x} = \frac{\int_0^b xy^2\,dx}{\int_0^b y^2\,dx},$$

$$= \frac{\int_0^b 4ax^2\,dx}{\int_0^b 4ax\,dx},$$

$$= \frac{\left[x^3/3\right]_0^b}{\left[x^2/2\right]_0^b},$$

$$= \frac{2}{3}b.$$

EXERCISE 29

1. Find the moment of inertia of a rectangle of sides a and b, about the side of length a.

2. Find the moment of inertia about its axis of a cylinder of radius r and height h.

3. Find the moment of inertia about its base of a triangle of base a and height h. (It is enough to consider the moment of inertia of a right angled triangle.)

4. Find the moment of inertia about its axis of a cone of height h and base radius r.

5. Find the moment of inertia of a ring of radius r, about a diameter.

6. Find the moment of inertia of a circle of radius r, about a diameter.

7. Find the moment of inertia of a sphere of radius r, about a diameter.

8. Find the moment of inertia of the solid obtained by revolving the ellipse $x^2/a^2+y^2/b^2 = 1$ about the x-axis, about its major axis.

9. Find the moment of inertia of an arc of the catenary $y = c \cosh x/c$, for $|x| \leqslant c$, about the y-axis.

10. Find the centre of gravity (c.g.) of a semi-circular arc of radius r.

11. Find the c.g. of a semicircle of radius r.

12. Find the c.g. of a hemisphere of radius r.

13. Find the c.g. of a cone of height h and base radius r.

14. Find the c.g. of the paraboloid of revolution of height h and base radius r. (Consider the rotation of the parabola $hy^2 = r^2x$ about the x-axis.)
15. Find the c.g. of the arc of the catenary $y = c \cosh x/c$ defined by $|x| \leqslant c$.
16. Find the c.g. of the area bounded by the cycloid $x = a(t - \sin t)$, $y = a(1 - \cos t)$, for $0 \leqslant t \leqslant 2\pi$, and the x-axis.
17. Find the c.g. of the solid generated by rotating the cardioid $r = a(1 + \cos \theta)$, $0 \leqslant \theta \leqslant \pi$, about the x-axis. (Express parametrically in cartesian coordinates.)
18. Find the c.g. of the area bounded by the cissoid $x = 2a \sin^2 t$, $y = 2a \sin^2 t \tan t$ for $\pi/12 \leqslant t \leqslant \pi/6$, and the x-axis.
19. Find the c.g. of the area bounded by the astroid $x = a \cos^3 t$, $y = a \sin^3 t$, for $0 \leqslant t \leqslant \pi/2$.
20. Find the c.g. of the area bounded by the curve $y = a \cos \dfrac{2\pi}{b} x$, the x-axis and the y-axis.
21. Find the c.g. of the solid generated by revolving the cycloid $x = a(t - \sin t)$, $y = a(1 - \cos t)$ about the x-axis $0 \leqslant t \leqslant \pi$.
22. Find the c.g. of the area of the loop of the curve $x = t^2, y = t - t^3/3$.
23. Find the c.g. of the arc of the curve $x = t^2$, $y = t - t^3/3$, defined by the range $0 \leqslant t \leqslant \sqrt{3}$.
24. Find the c.g. of the area bounded by the curve $x = t^2 - t$, $y = t^3 + t^2$, and the x-axis.
25. Find the c.g. of the area bounded by the parabolas $y^2 = ax$ and $x^2 = ay$.
26. Find the c.g. of the area of the cardioid $r = a(1 + \cos \theta)$.
27. Find the c.g. of the area bounded by the parabola $y^2 = 4ax$, the x-axis and the line $x = x_0$.

§ 4. Irregular integrals

In this section we are concerned with the integration of functions which are not defined or are not bounded at a finite number of points, and with integrals in which the interval of integration is infinitely large.

4.1 Integrals of functions not defined at a finite number of points

Let the function $f(x)$ be defined and bounded in the closed interval $[a, b]$, except at a finite number of points $x_1, x_2, \ldots, x_k$. We define arbitrarily the value of the function $f(x)$ at these points, as $\alpha_1, \alpha_2, \ldots, \alpha_k$. We then define a function $\phi(x)$ as follows,

$$\phi(x) = \begin{cases} f(x) & (x \neq x_i, \quad i = 1, 2, \ldots, k) \\ \alpha_i & (x = x_i, \quad i = 1, 2, \ldots, k) \end{cases}.$$

If $\phi(x)$ is integrable in $[a, b]$ then

$$\int_a^b \phi(x)\, dx$$

is the irregular integral of $f(x)$ in this interval. If we define $f(x_1), f(x_2), \ldots, f(x_k)$ in a different way we obtain another function $\phi_1(x)$. However,

$$\int_a^b \{\phi(x) - \phi_1(x)\}\, dx = 0,$$

since $\phi(x) - \phi_1(x)$ is zero everywhere except at a finite number of points. Thus if $\phi(x)$ is integrable so also is $\phi_1(x)$ and

$$\int_a^b \phi(x)\, dx = \int_a^b \phi_1(x)\, dx.$$

Consequently the irregular integral does not depend on the way in which we define $f(x)$ at the given points. We will therefore denote the irregular integral in the same way as usual, namely

$$\int_a^b f(x)\, dx.$$

For example, if $f(x) = \sin \frac{1}{x}$, $x \neq 0$, we can define

$$\phi(x) = \begin{cases} f(x), & x \neq 0 \\ 0, & x = 0 \end{cases}.$$

which is integrable in any interval since it is only discontinuous at the point $x = 0$. Thus

$$\int_{-1}^1 \sin \frac{1}{x}\, dx = \int_{-1}^1 \phi(x)\, dx.$$

4.2 *Integrals of unbounded functions*

We assume that $f(x)$ is defined in the interval $[a, b]$, except perhaps at the point $x = a$ where it is unbounded. If $f(x)$ is integrable in any interval $(a+\varepsilon, b)$, where $\varepsilon > 0$, $a+\varepsilon < b$, and

$$\lim_{\varepsilon \to 0^+} \int_{a+\varepsilon}^{b} f(x)\, dx \quad \text{exists,}$$

then the limit is the irregular integral

$$\int_{a}^{b} f(x)\, dx.$$

4.3 *Integrals of infinite interval*

Let $f(x)$ be defined for all $x > a$. If in any interval $a \leqslant x \leqslant b$ the function is integrable and if

$$\lim_{b \to \infty} \int_{a}^{b} f(x)\, dx \quad \text{exists,}$$

then we call this limit the irregular integral of $f(x)$ from a to ∞, and we write

$$\int_{a}^{\infty} f(x)\, dx.$$

We say that this integral is convergent. If the limit does not exist we call the integral divergent.

4.4 *Geometric interpretation*

(i) If $f(x)$ is non-negative in $a \leqslant x \leqslant b$, if $\lim\limits_{x \to b^-} f(x) \to +\infty$, and

$$\lim_{\varepsilon \to 0^+} \int_{a}^{b-\varepsilon} f(x)\, dx \quad \text{exists,}$$

then we say that the area bounded by the lines $x = a$, the

x-axis, $y = f(x)$ and $x = b$, (which is a finite area with an infinite bound) (Fig. 83), is equal to

$$\int_a^b f(x)\,\mathrm{d}x.$$

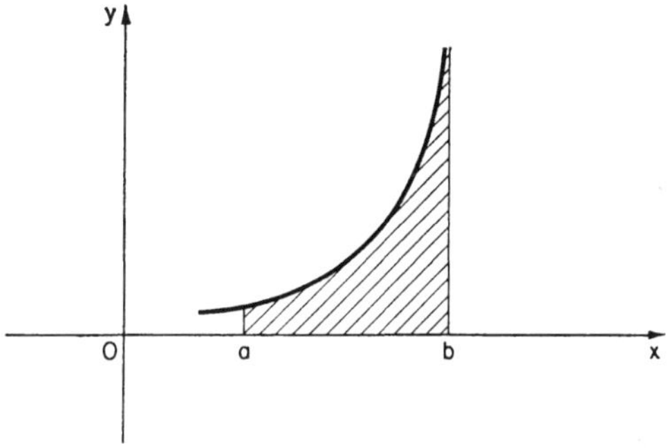

FIG. 83

(ii) If $f(x)$ is continuous for $x > 0$, and if

$$\int_a^\infty |f(x)|\,\mathrm{d}x \quad \text{exists,}$$

then we say that the integral expresses the area bounded by $x = a$, the curve $y = f(x)$ and the x-axis to the right of $x = a$ (which is an unlimited bound) (Fig. 84).

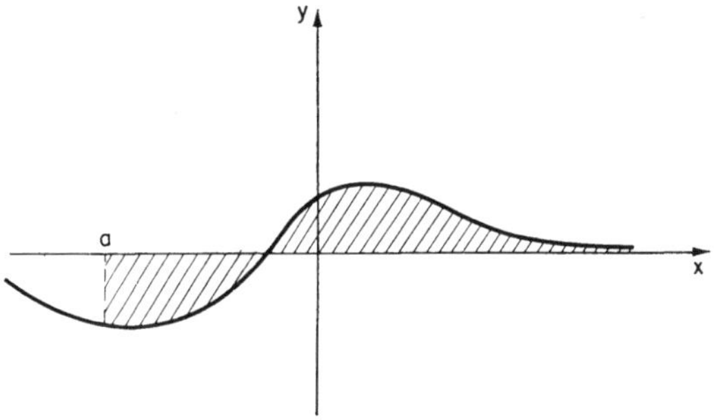

FIG. 84

Examples

1. Find the value of $\int_0^3 \frac{1}{\sqrt{x}}\,dx$.

Solution. The integrand is not continuous at $x = 0$. For $\varepsilon > 0$ we calculate

$$\int_\varepsilon^3 \frac{1}{\sqrt{x}}\,dx = \Big[2\sqrt{x}\Big]_\varepsilon^3,$$
$$= 2\sqrt{3} - 2\sqrt{\varepsilon},$$
$$\to 2/3 \text{ as } \varepsilon \to 0^+.$$

Thus,
$$\int_0^3 \frac{1}{\sqrt{x}}\,dx = \lim_{\varepsilon \to 0^+} \int_\varepsilon^3 \frac{1}{\sqrt{x}}\,dx,$$
$$= 2\sqrt{3}.$$

2. Find the area bounded by the lines $x = 0$, $x = 9$, the x-axis and the curve

$$y = \frac{1}{\sqrt[3]{(x-1)}} \qquad \text{(Fig. 85).}$$

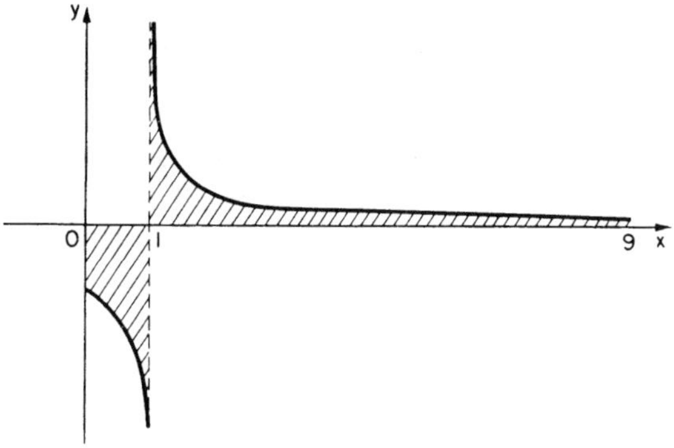

FIG. 85

Solution. This function is discontinuous for $x = 1$, a point which lies within the interval [0, 9]. For $x < 1$, $f(x) < 0$, for

$x > 1, f(x) > 0$. Thus the required area is

$$A = \int_0^9 \frac{1}{|\sqrt[3]{(x-1)}|}\,dx,$$

$$= -\lim_{\varepsilon \to 0^+} \int_0^{1-\varepsilon} f(x)\,dx + \lim_{\varepsilon' \to 0^+} \int_{1+\varepsilon'}^9 f(x)\,dx,$$

$$= -\lim_{\varepsilon \to 0^+} \left[\frac{3}{2}(x-1)^{2/3}\right]_0^{1-\varepsilon} + \lim_{\varepsilon' \to 0^+} \left[\frac{3}{2}(x-1)^{2/3}\right]_{1+\varepsilon'}^9,$$

$$= -\lim_{\varepsilon \to 0^+} \left\{\frac{3}{2}(-\varepsilon)^{2/3} - \frac{3}{2}(-1)^{2/3}\right\}$$

$$+ \lim_{\varepsilon' \to 0^+} \left\{6 - \frac{3}{2}(\varepsilon')^{2/3}\right\},$$

$$= +\frac{3}{2} + 6,$$

$$= \frac{15}{2}.$$

3. Find the value of $\int_0^1 \frac{1}{x\sqrt{x}}\,dx$.

Solution. The integrand is continuous in the given range except at $x = 0$. Then provided the limit exists,

$$\int_0^1 \frac{1}{x\sqrt{x}}\,dx = \lim_{\varepsilon \to 0^+} \int_\varepsilon^1 \frac{1}{x\sqrt{x}}\,dx$$

$$= \lim_{\varepsilon \to 0^+} \left[-\frac{2}{\sqrt{x}}\right]_\varepsilon^1$$

$$= \lim_{\varepsilon \to 0^+} \left(-2 + \frac{2}{\varepsilon}\right)$$

$$\to \infty.$$

4. Calculate $\int_1^\infty \left(\frac{2}{x} + \frac{1}{x^2}\right)^2 dx.$

Solution. Provided the limit exists

$$\int_1^\infty \left(\frac{2}{x}+\frac{1}{x^2}\right)^2 dx = \lim_{K\to\infty} \int_1^K \left(\frac{2}{x}+\frac{1}{x^2}\right)^2 dx,$$

$$= \lim_{K\to\infty} \left[-\frac{4}{x}-\frac{2}{x^2}-\frac{1}{3x^3}\right]_1^K,$$

$$= \lim_{K\to\infty} \left\{-\frac{4}{K}-\frac{2}{K^2}-\frac{1}{3K^3}-\left(-4-2-\frac{1}{3}\right)\right\},$$

$$= \frac{19}{3}.$$

5. Find the area bounded by the x-axis and the curve

$$y = \frac{x}{x^4+1} \quad \text{(Fig. 86).}$$

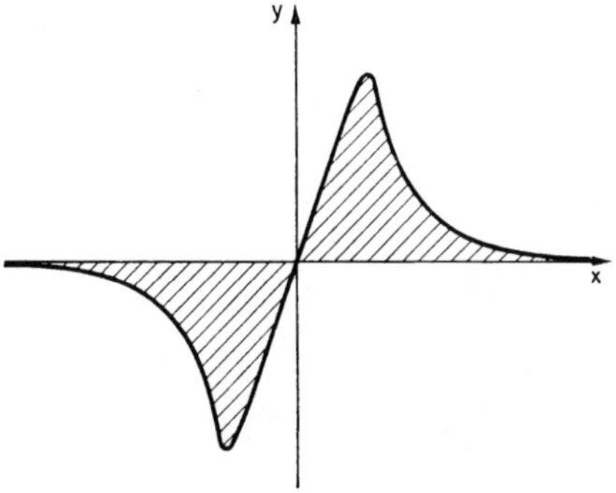

FIG. 86

Solution. The required area is

$$A = \int_{-\infty}^\infty \left|\frac{x}{x^4+1}\right| dx$$

$$= -\int_{-\infty}^0 \frac{x}{x^4+1} dx + \int_0^\infty \frac{x}{x^4+1} dx$$

$$= 2\int_0^\infty \frac{x}{x^4+1}\,dx$$

$$= 2\lim_{K\to\infty}\int_0^K \frac{x}{x^4+1}\,dx$$

$$= 2\lim_{K\to\infty}\left[\frac{1}{2}\arctan x^2\right]_0^K$$

$$= \lim_{K\to\infty}(\arctan K^2)$$

$$= \frac{1}{2}\pi.$$

6. Calculate $\int_0^\infty e^{-x}\,dx.$

If the limit exists,

$$\int_0^\infty e^{-x}\,dx = \lim_{K\to\infty}\int_0^K e^{-x}\,dx$$

$$= \lim_{K\to\infty}\Big[-e^{-x}\Big]_0^K$$

$$= \lim_{K\to\infty}(-e^{-K}+1)$$

$$= 1.$$

7. Calculate $$\int_{-\infty}^{\sqrt{3}} \frac{dx}{x^2+1}.$$

Solution. If the limit exists,

$$\int_{-\infty}^{\sqrt{3}} \frac{dx}{x^2+1} = \lim_{K\to\infty}\int_{-K}^{\sqrt{3}} \frac{dx}{x^2+1}$$

$$= \lim_{K\to\infty}\Big[\arctan x\Big]_{-K}^{\sqrt{3}}$$

$$= \lim_{K\to\infty}\left\{\frac{1}{3}\pi - \arctan(-K)\right\}$$

$$= \frac{1}{3}\pi - \left(-\frac{1}{2}\pi\right)$$

$$= \frac{5}{6}\pi.$$

EXERCISE 30

Find the values of the following:

1. $\int_0^1 \frac{x}{1-x}\,dx.$

2. $\int_0^1 x^{-1/3}\,dx.$

3. $\int_0^2 x^{-4/3}\,dx.$

4. $\int_0^{16} x^{-3/4}\,dx.$

5. $\int_0^{16} (4x^3)^{-1/4}\,dx.$

6. $\int_0^1 x^{-3/5}\,dx.$

7. $\int_0^a \frac{dx}{\sqrt{(a^2-x^2)}}.$

8. $\int_a^b \frac{dx}{\sqrt{\{(x-a)(b-x)\}}},\quad a < b.$

9. $\int_0^{\sqrt{(2/3)}} \frac{x}{\sqrt{(4-9x^4)}}\,dx.$

10. $\int_0^1 \frac{x}{\sqrt{(1-x^2)}}\,dx.$

11. $\int_{-2}^0 \frac{1}{(2-x)^2}\cdot\left(\frac{2-x}{2+x}\right)^{1/3}\,dx.$

12. $\int_1^2 \frac{dx}{\sqrt{(x^2-1)}}.$

13. $\int_2^3 \frac{x}{\sqrt{(x^2-4)}}\,dx.$

14. $\int_0^6 \frac{2x}{(x^2-4)^{2/3}}\,dx.$

15. $\int_0^1 \frac{x^2}{\sqrt{(x-x^2)}}\,dx.$

16. $\int_1^2 \frac{dx}{x\sqrt{(x^2-1)}}.$

17. $\int_{-2}^{-1} \frac{dx}{x\sqrt{(x^2-1)}}.$

18. $\int_0^1 \frac{dx}{\sqrt{(4x-4x^2)}}.$

19. $\int_0^1 \frac{\sqrt{x}}{\sqrt{(1-x)}}\,dx.$

20. $\int_2^4 \frac{x\sqrt{(x-2)}}{\sqrt{(4-x)}}\,dx.$

21. $\int_{1/2}^1 \frac{dx}{\sqrt{(1-x^2}\arcsin x)}.$

22. $\int_0^{\pi/4} \sec^2(2x)\,dx.$

23. $\int_0^{\pi/2} \tan x\,dx.$

24. $\int_{-\infty}^{+\infty} \frac{dx}{1+x^2}.$

25. $\int_{-\infty}^{+\infty} \frac{(\arctan x)^2}{1+x^2}\,dx.$

26. $\int_{\sqrt{3}}^{\infty} \frac{dx}{x^2+9}.$

27. $\int_3^{\infty} \frac{dx}{x^2}.$

28. $\int_4^\infty \frac{dx}{(x-3)^2}$.

29. $\int_1^\infty \frac{dx}{2x^2+4}$.

30. $\int_1^\infty \frac{dx}{x^2-6x+13}$.

31. $\int_{-\infty}^{-1/2} \frac{dx}{x^2+x+1}$.

32. $\int_{-\infty}^{+\infty} \frac{dx}{x^2+2x+2}$.

33. $\int_1^\infty \frac{dx}{x(x+1)^2}$.

34. $\int_0^\infty \frac{dx}{1+x^3}$.

35. $\int_0^\infty \frac{7x+2}{x^3-5x^2+12x-60}\,dx$.

36. $\int_1^\infty \frac{dx}{x\sqrt{(x^2-1)}}$.

37. $\int_0^\infty e^{-ax}\,dx, \quad a>0$.

38. $\int_0^\infty xe^{-x^2}\,dx$.

39. $\int_0^\infty e^{-x}\sin x\,dx$.

40. $\int_1^\infty \frac{e^{2/x}}{2x^2}\,dx$.

41. $\int_0^\pi \frac{dx}{a+b\cos x}, \quad a>b>0$.

42. $\int_0^{\pi/2} \frac{dx}{a^2\cos^2 x+b^2\sin^2 x}, \quad a>0, \quad b>0$.

CHAPTER 12

MAXIMA AND MINIMA OF A FUNCTION OF TWO VARIABLES

§ 1. Definition

Given a function $z=f(x, y)$, we say that the function has a maximum at the point (x_0, y_0) if there exists a neighbourhood of this point such that for any (x_0+h, y_0+k) belonging to the neighbourhood,

$$f(x_0+h,\ y_0+k) < f(x_0, y_0),$$

where h and k are not both zero.

There is a minimum at the point (x_0, y_0) if, under similar conditions,

$$f(x_0+h, y_0+k) > f(x_0, y_0).$$

§ 2. Theorem

Let a function of two variables, $z=f(x, y)$ have in the neighbourhood of $M(x_0, y_0)$ continuous partial first and second derivatives, where

$$f_x(x_0, y_0) = 0, \quad \text{and} \quad f_y(x_0, y_0) = 0.$$

We define

$$\Delta = f_{xx}(x_0, y_0) \cdot f_{yy}(x_0, y_0) - \{f_{xy}(x_0, y_0)\}^2.$$

Then,

(i) If $\Delta > 0$, the point M is an extreme point of the function.

If $f_{xx}(x_0, y_0) < 0$, it is a maximum.

If $f_{xx}(x_0, y_0) > 0$, it is a minimum.

Similarly, it is a maximum or minimum according to whether $f_{yy}(x_0, y_0)$ is negative or positive. Note that f_{xx} and f_{yy} are unequal to zero and must have the same sign, otherwise $\Delta < 0$.

(ii) If $\Delta < 0$, the point M is not an extreme point of the function.

(iii) If $\Delta = 0$, the point M may or may not be an extreme point of the function. For example, consider the two functions

$$z = x^4 + y^4, \quad \text{and} \quad w = x^3 + y^3.$$

For both functions $\Delta = 0$ at the origin, but $z(0, 0)$ is a minimum, while $w(0, 0)$ is not an extreme point.

Examples

1. Find the extreme points of the function

$$f(x, y) = 3x^3 + 3x^2y - y^3 - 15x.$$

Solution

$$f_x = 9x^2 + 6xy - 15, \qquad f_y = 3x^2 - 3y^2,$$
$$f_{xx} = 18x + 6y, \qquad f_{yy} = -6y,$$
$$f_{xy} = 6x.$$

For the extreme points

$$f_x = 0, \quad \text{and} \quad f_y = 0,$$

which gives

$$3x^2 + 2xy - 5 = 0, \quad \text{and} \quad x^2 - y^2 = 0.$$

Solving this pair of simultaneous equations, we obtain the four points

$$A(1, 1), \quad B(-1, -1), \quad C(\sqrt{5}, -\sqrt{5}), \quad D(-\sqrt{5}, \sqrt{5}).$$

Tabulating the values of the second order derivatives and of Δ at these points, we obtain:

	f_{xx}	f_{xy}	f_{yy}	$\Delta = f_{xx}f_{yy} - (f_{xy})^2$	Conclusion
A	24	6	-6	$-24\times 6-6^2<0$	No extreme
B	-24	-6	$+6$	$-24\times 6-6^2<0$	No extreme
C	$12\sqrt{5}$	$6\sqrt{5}$	$6\sqrt{5}$	$72\times 5-36\times 5>0$	Minimum ($f_{xx}>0$)
D	$-12\sqrt{5}$	$-6\sqrt{5}$	$-6\sqrt{5}$	$72\times 5-36\times 5>0$	Maximum ($f_{xx}<0$)

2. Find the coordinates of the points on the parabola $y^2 = 6x$ such that the distance from $A(3, 12)$ is either a maximum or a minimum.

Solution. The square of the distance from A to any point $P(x, y)$ on the parabola is

$$z = (x-3)^2+(y-12)^2.$$

We should notice that this is not strictly a function of two variables since in addition y is a function of x. In other words we can obtain z as a function of one variable. Thus

$$z = (y^2/6-3)^2+(y-12)^2,$$
$$= y^4/36-24y+153.$$

Then, $\dfrac{dz}{dy} = \dfrac{1}{9}y^3-24$, and $\dfrac{d^2z}{dy^2} = \dfrac{1}{3}y^2$.

Therefore $\dfrac{dz}{dy} = 0$ for $y = 6$, for which value $\dfrac{d^2z}{dy^2} > 0$. Thus the point on the parabola for which z is a minimum is (6, 6), the minimum distance being $3\sqrt{5}$.

3. Find the dimensions of the rectangular open tank of volume 32 m^3 such that the surface area is a minimum.

Solution. Let the dimensions of the tank be x, y, z (Fig. 87). Then,

$$xyz = 32,$$

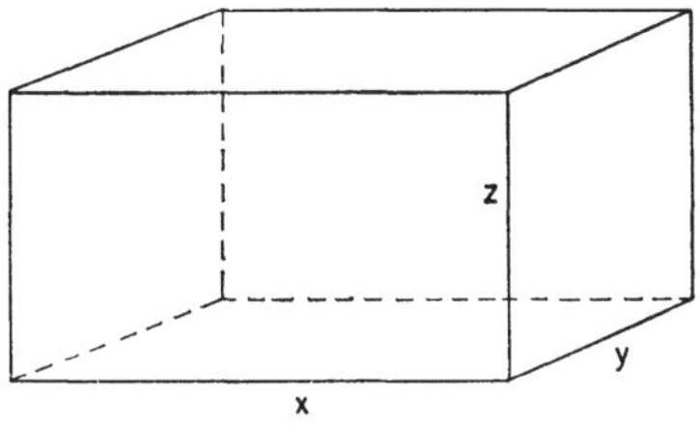

FIG. 87

and the surface area S is given by

$$S = 2yz+2zx+xy,$$

which as a function of x and y is

$$S = \frac{64}{x}+\frac{64}{y}+xy.$$

We now differentiate to obtain the partial derivatives,

$$\frac{\partial S}{\partial x} = -\frac{64}{x^2}+y, \qquad \frac{\partial S}{\partial y} = -\frac{64}{y^2}+x,$$

$$\frac{\partial^2 S}{\partial x^2} = \frac{128}{x^3}, \qquad \frac{\partial^2 S}{\partial x \partial y} = 1, \qquad \frac{\partial^2 S}{\partial y^2} = \frac{128}{y^3}.$$

At the extreme points of S the first order partial derivatives are zero, and the points therefore satisfy the equations

$$64-x^2y = 0, \quad \text{and} \quad 64-y^2x = 0.$$

The solution of these equations is $x = y$, and hence

$$64-x^3 = 0,$$

giving the point $x = 4$, $y = 4$ as a possible extreme point of the function. At this point the values of the partial derivatives are as follows,

$$\frac{\partial^2 S}{\partial x^2} = 2, \qquad \frac{\partial^2 S}{\partial x \partial y} = 1, \qquad \frac{\partial^2 S}{\partial y^2} = 2,$$

thus,

$$\Delta = \frac{\partial^2 S}{\partial x^2} \cdot \frac{\partial^2 S}{\partial y^2} - \left\{ \frac{\partial^2 S}{\partial x \partial y} \right\}^2,$$
$$= 4 - 1 > 0.$$

Thus the point (4, 4) is a minimum point of the function and the dimensions of the tank which give a minimum surface area are $x = 4$ m, $y = 4$ m, $z = 2$ m.

EXERCISE 31

Find the extreme values of the function $z = f(x, y)$, where

1. $f(x, y) = 2x^2 + 3xy + y^2 - 2x - y + 1$.
2. $f(x, y) = x^4 + y^4 - 2x^2 + 4xy - 2y^2$.
3. $f(x, y) = x^2 - xy + 2y^2 - x + 4y - 5$.
4. $f(x, y) = x^3 + y^3 - 3axy$.
5. $f(x, y) = x^3 + 3x^2y - 6xy - 3y^2 - 15x - 15y$.
6. $f(x, y) = x^2 - xy + y^2 + 3x - 2y + 1$.
7. $f(x, y) = x^2 + y^2 + xy - 6x - 4y + 5$.
8. $f(x, y) = x^2 - 6xy + y^3 + 3x + 6y$.
9. $f(x, y) = x^3 - 3axy + y^3$.
10. $f(x, y) = 4xy + \dfrac{1}{x} + \dfrac{1}{y}$.
11. $f(x, y) = x^2 - xy + y^2 - 2x + y$.
12. $f(x, y) = x^3 + y^2 - 6xy - 48x$.
13. $f(x, y) = (x + y)^2 - (x + 5y + xy)$.
14. $f(x, y) = 1 - \sqrt{(x^2 + y^2)}$.
15. $f(x, y) = xy^2(a - x - y)^3$.
16. $f(x, y) = (6 - x - y)x^2y^3$.

17. $f(x, y) = \sin x+\sin y+\sin (x+y)$.

18. $f(x, y) = \sin x+\cos y+\cos(x-y)$.

19. $f(x, y) = \sin x \sin y \sin (a-x-y)$.

20. $f(x, y) = \sin x \sin y \sin(x+y), \quad 0 \leqslant x \leqslant \pi, \quad 0 \leqslant y \leqslant \pi$.

21. $f(x, y) = x-2y+\log \sqrt{(x^2+y^2)}+3 \text{ arc tan } \dfrac{y}{x}$.

22. $f(x, y) = x^2+xy+y^2-4 \log |x|-10 \log |y|$.

23. $f(x, y) = (8x^2-6xy+3y^2)e^{2x+3y}$.

24. $f(x, y) = (5x+7y-25) \exp\{-(x^2+xy+y^2)\}$.

25. Find the maximum area of a triangle which can be inscribed in a circle.

26. Divide a given number n into three parts so that the product of the parts is a maximum.

27. Find the maximum area of a triangle with given semiperimeter s.

28. Find the coordinates of a point on the circle $(x-6)^2+(y-1)^2 = 25$, such that the distance of the point from $P(0, 7)$ is a maximum or a minimum.

29. Find the coordinates of the points on the parabola $2y-x^2 = 0$ whose distance from $P(4, 1)$ is a maximum or a minimum.

30. Find the dimensions of the cuboid of maximum volume which can be circumscribed by a sphere of radius r.

CHAPTER 13

TANGENTS AND NORMALS. CURVATURE

§ 1. Tangents and normals

1.1 Equation of tangent and normal

(i) If a curve is given by the equation $y = f(x)$ where $f(x)$ is differentiable, then the equation of the tangent at the point $P(x_0, y_0)$ on the curve is

$$y - y_0 = f'(x_0)(x - x_0),$$

and the equation of the normal is

$$(x - x_0) + f'(x_0)(y - y_0) = 0,$$

since the normal to the curve at the point P of the curve is the perpendicular to the tangent at P.

(ii) Given a curve $F(x, y) = 0$ and a point (x_0, y_0) lying on the curve. If the partial first derivatives $F_x(x_0, y_0)$ and $F_y(x_0, y_0)$ exist and at least one of these is not equal to zero, and if there exists a neighbourhood of (x_0, y_0) such that inside this neighbourhood both derivatives are continuous, then the curve has a tangent at (x_0, y_0), given by the equation

$$F_x(x_0, y_0)\,(x - x_0) + F_y(x_0, y_0)(y - y_0) = 0,$$

and there also exists a normal at the point with equation

$$F_y(x_0, y_0)(x - x_0) - F_x(x_0, y_0)(y - y_0) = 0.$$

(The symbols $F_x(x_0, y_0)$ and $F_y(x_0, y_0)$ denote the values of the partial derivatives $\partial F/\partial x$, $\partial F/\partial y$ at the point (x_0, y_0).)

Examples

1. Find the equations of the tangent and normal to the curve $y = x^3 - x$ at the point $P(1, 0)$.

Solution. First we verify that the point P lies on the curve. Right-hand side $= 1-1 = 0 =$ Left-hand side.
Next we find the value of the first derivative at P.

$$\begin{aligned} y' &= 3x^2-1, \\ y'(1) &= 3-1 \\ &= 2. \end{aligned}$$

Then the equation of the tangent is

$$y = 2(x-1), \quad \text{or} \quad y = 2x-2,$$

and the equation of the normal is

$$(x-1)+2y = 0, \quad \text{or} \quad y = -\frac{1}{2}x+\frac{1}{2}.$$

2. Find the equations of the tangent and normal to the curve $F(x, y) \equiv x^2+y^3 = 0$ at the point $P(1, -1)$.

Solution. First we verify that the point P lies on the curve. Left-hand side $= 1^2+(-1)^3 = 1-1 = 0 =$ Right-hand side. Next we find the values of the partial derivatives at P.

$$\begin{aligned} F_x &= 2x, & F_x(1,\ -1) &= 2; \\ F_y &= 3y^2, & F_y(1,\ -1) &= 3. \end{aligned}$$

Then the equation of the tangent is

$$2(x-1)+3(y+1) = 0, \quad \text{or} \quad 2x+3y+1 = 0,$$

and the equation of the normal is

$$3(x-1)-2(y+1) = 0, \quad \text{or} \quad 3x-2y-5 = 0.$$

3. Find the equations of the tangent and normal to the cycloid $x = a(t-\sin t)$, $y = a(1-\cos t)$, $a > 0$, at the point $t = \frac{1}{3}\pi$.

Solution. Since all real values of t determine a point of the cycloid it is unnecessary to verify that the point lies on the curve. By differentiating the parametric relationships we can obtain the first derivative of y with respect to x.

$$dx = a(1-\cos t)\,dt, \qquad dy = a\sin t\,dt,$$

hence, assuming $\cos t \neq 1$,

$$\frac{dy}{dx} = \frac{\sin t}{1-\cos t}.$$

At the point $t = \frac{1}{3}\pi$ we obtain

$$x_0 = a\left(\frac{1}{3}\pi - \frac{\sqrt{3}}{2}\right), \quad y_0 = \frac{1}{2}a,$$

$$\left(\frac{dy}{dx}\right)_{(t=\pi/3)} = \frac{\sqrt{3}/2}{1-1/2} = \sqrt{3}.$$

Therefore the equation of the tangent at the given point is

$$y - \frac{1}{2}a = \sqrt{3}\left\{x - a\left(\frac{1}{3}\pi - \frac{\sqrt{3}}{2}\right)\right\},$$

or

$$y = x\sqrt{3} + a\left(2 - \frac{\pi}{\sqrt{3}}\right).$$

The equation of the normal is

$$x - a\left(\frac{1}{3}\pi - \frac{\sqrt{3}}{2}\right) + \sqrt{3}\left(y - \frac{1}{2}a\right) = 0,$$

or

$$x + y\sqrt{3} - \frac{1}{3}\pi a = 0.$$

EXERCISE 32

1. Find the equations of the tangent and normal to the curve $y = (2x^3 + x^2 + 1)^{1/2}$ at the point $P(1, 2)$.
2. Find the equations of the tangent and normal to the curve $y = x^3 - 2x + 1$ at the point $P(1, 0)$.
3. Find the equations of the tangent and normal to the curve $x^2 - 2xy + y^2 - x + 1 = 0$ at the point $P(2, 1)$.
4. Find the equations of the tangent and normal to the curve $x^2y^3 - y^2 - 4 = 0$ at the point $P(1, 2)$.
5. Find the equations of the tangent and normal to the curve whose parametric form is $x = 3t^2$, $y = 3t - t^3$ at the point $t = 1/\sqrt{3}$.
6. Find the equation of the tangent to the curve $x^3 - py^2 = 0$ (a semi-cubical parabola) at the point (p, p).
7. Find the equation of the tangent to the cissoid $y^2(2r - x) - x^3 = 0$ at the point (r, r).
8. Find the equation of the tangent to the curve $x^3 - 3axy + y^3 = 0$ at the point $(3a/2, 3a/2)$.
9. Find the equation of the tangent to the curve $y = a^x$ at the point $(1, a)$.
10. Find the equation of the tangent to the curve $x^3 + y^3 + 2x - 6 = 0$ at the point of the curve whose abscissa is -3.
11. Find the equation of the tangent to the catenary $y = c \cosh x/c$ at the point of the curve whose x coordinate is c.
12. Prove that the segment of the tangent to the astroid $x^{2/3} + y^{2/3} = a^{2/3}$ between the x- and y-axes is of constant length a.
13. Find the equation of the normal to the curve $x^3 + y^2 + 2x - 6 = 0$ at the point where $x = 3$.
14. Find the equation of the normal to the curve $y = a \log \cos x/a$ at the point where $x = 2\pi a$.
15. Find the equations of the tangent and normal to the curve $r = e^\theta$ (a logarithmic spiral), at the point $\theta = k\pi$, where k is an integer. (Put $x = r \cos \theta$, $y = r \sin \theta$).
16. Find the equations of the tangent and normal to the curve $x = r(\cos t + t \sin t)$, $y = r(\sin t - t \cos t)$, $r > 0$, at the point $t = \pi/4$.

§ 2. Curvature and radius of curvature

2.1 Definition of curvature

The mean curvature $\varkappa(PP')$ of the arc PP' is defined as the ratio of the angle between the tangents to the curve at P and P' to the length of the arc PP' (Fig. 88), that is,

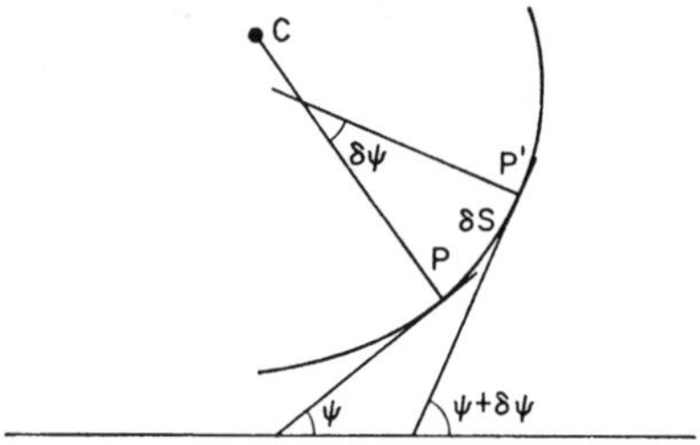

FIG. 88

$\varkappa(PP') = \delta\psi/\delta s$, where the arc length PP' is δs, and where the angle ψ is measured in an anticlockwise sense from the positive direction of the x-axis.

The curvature $\varkappa(P)$ at P is defined as the limit of the mean curvature of the arc PP' as P' tends to P along the curve, that is,

$$\varkappa(P) = \lim_{\delta s \to 0} \frac{\delta\psi}{\delta s}$$

$$= \frac{d\psi}{ds}.$$

Thus $\varkappa$ will be positive or negative according as ψ increases or decreases with s.

2.2 *Definition of radius of curvature*

If the point of intersection of the normals to the curve at P and P' tends to a point C as P' tends to P along the curve, then the length CP is defined as the radius of curvature (ϱ) of the curve at the point P (Fig. 88).

$$\frac{1}{\varrho} = \lim_{\delta s \to 0} \frac{\delta\psi}{\delta s},$$

$$= \frac{d\psi}{ds}.$$

Thus, $$\varkappa = \frac{1}{\varrho}.$$

A curve is called strictly tangential to another curve at a common point of the two curves if at this point each has the same centre of curvature and the same radius of curvature.

2.3 Formulae for curvature and radius of curvature

(i) Given a curve $y = f(x)$, then the curvature $\varkappa$ at a point of the curve at which the second derivative exists is given by

$$\varkappa = \frac{\mathrm{d}^2y/\mathrm{d}x^2}{\{1+(\mathrm{d}y/\mathrm{d}x)^2\}^{3/2}}, \quad \text{or} \quad \varkappa = \frac{f''(x)}{[1+\{f'(x)\}^2]^{3/2}},$$

and $$\varrho = \frac{\{1+(\mathrm{d}y/\mathrm{d}x)^2\}^{3/2}}{\mathrm{d}^2y/\mathrm{d}x^2}, \quad \text{or} \quad \varrho = \frac{[1+\{f'(x)\}^2]^{3/2}}{f''(x)}.$$

Note:

(a) If $y = f(x)$ has a turning point at $P(x_0, y_0)$, then since $f'(x_0) = 0$,

$$\varkappa(P) = f''(x_0).$$

(b) If $y = f(x)$ has a point of inflexion at $Q(x_1, y_1)$, then since $f''(x_1) = 0$,

$$\varkappa(Q) = 0.$$

(ii) Given a curve in parametric form $x = x(t)$, $y = y(t)$, at a point of the curve at which the second derivatives exist the curvature $\varkappa$ is given by

$$\varkappa = \frac{\dot{x}\ddot{y} - \ddot{x}\dot{y}}{(\dot{x}^2 + \dot{y}^2)^{3/2}},$$

and the radius of curvature ϱ is correspondingly

$$\varrho = \frac{(\dot{x}^2 + \dot{y}^2)^{3/2}}{\dot{x}\ddot{y} - \ddot{x}\dot{y}}.$$

2.4 Sign convention

From the definition of curvature as

$$\varkappa = \frac{d\psi}{ds},$$

it follows that the curvature will be positive or negative according as ψ increases or decreases with s. We always take the positive sign of $\{1+(dy/dx)^2\}^{3/2}$ and $(\dot{x}^2+\dot{y}^2)^{3/2}$, and consequently the sign of $\varkappa$ is the same as the sign of d^2y/dx^2 and $(\dot{x}\ddot{y}-\ddot{x}\dot{y})$. Thus $\varkappa$ is positive when the curve is concave upwards and negative when the curve is concave downwards. The sign of $\varkappa$ determines the sign of ϱ.

Examples

1. Find the radius of curvature of the parabola $y = ax^2 + bx + c$, $(a \neq 0)$, at the point $x = -\dfrac{b}{2a}$.

Solution. From the equation we obtain

$$y' = 2ax+b, \qquad y'' = 2a,$$

therefore,

$$\varrho = \frac{\left\{1+\left[2a\left(-\dfrac{b}{2a}\right)+b\right]^2\right\}^{3/2}}{2a},$$

that is,

$$\varrho = \frac{1}{2a}.$$

2. Given the curve $y = \log x$, find the least value of the curvature.

Solution. Assuming $x > 0$, then

$$y' = \frac{1}{x}, \qquad y'' = -\frac{1}{x^2},$$

and $$(1+y'^2)^{3/2} = (1+1/x^2)^{3/2},$$
$$= \frac{(x^2+1)^{3/2}}{x^3},$$

therefore, $$\varkappa = -\frac{x}{(1+x^2)^{3/2}} \tag{1}$$

The turning points of this function are given by the zero values of $\frac{\mathrm{d}\varkappa}{\mathrm{d}x}$.

$$\frac{\mathrm{d}\varkappa}{\mathrm{d}x} = \frac{(1+x^2)^{3/2}(-1)-(-x)3/2(1+x^2)^{1/2}\cdot 2x}{(1+x^2)^3},$$
$$= \frac{-1-x^2+3x^2}{(1+x^2)^{5/2}},$$
$$= \frac{2x^2-1}{(1+x^2)^{5/2}}.$$

Hence $\frac{\mathrm{d}\varkappa}{\mathrm{d}x} = 0$ if $2x^2-1 = 0$, i.e. if $x = \pm 1/\sqrt{2}$. But $x > 0$, therefore $x = 1/\sqrt{2}$ is the only possible turning point. Moreover, as x increases through the value $1/\sqrt{2}$ the sign of $\mathrm{d}\varkappa/\mathrm{d}x$ changes from negative to positive, hence $\varkappa$ is a minimum for $x = 1/\sqrt{2}$, and substituting in (1) we obtain

$$\varkappa_{\min} = -\frac{2}{3\sqrt{3}}.$$

3. Find the smallest and greatest radius of curvature of the cardioid $r = a(1+\cos\theta)$, $a > 0$.

Solution. Transforming into cartesian coordinates:

$$x = a\cos\theta(1+\cos\theta), \qquad y = a\sin\theta(1+\cos\theta).$$

Since the cardioid is symmetrical about the x-axis it is sufficient to consider the interval $0 \leqslant \theta \leqslant \pi$.

Differentiating with respect to θ:

$$x' = -a(\sin\theta+\sin 2\theta), \qquad y' = a(\cos\theta+\cos 2\theta),$$
$$x'' = -a(\cos\theta+2\cos 2\theta), \qquad y'' = -a(\sin\theta+2\sin 2\theta).$$

Hence,

$$\begin{aligned} x'^2+y'^2 &= a^2(\sin^2\theta+2\sin\theta\sin 2\theta+\sin^2 2\theta+\cos^2\theta \\ &\qquad +2\cos\theta\cos 2\theta+\cos^2 2\theta), \\ &= a^2\{2+2\cos(2\theta-\theta)\}, \\ &= 2a^2(1+\cos\theta), \\ &= 4a^2\cos^2\frac{\theta}{2}. \end{aligned}$$

Also,

$$\begin{aligned} x'y''-x''y' &= a^2(\sin^2\theta+3\sin\theta\sin 2\theta+2\sin^2 2\theta \\ &\qquad +\cos^2\theta+3\cos\theta\cos 2\theta+2\cos^2 2\theta) \\ &= a^2\{3+3\cos(2\theta-\theta)\}, \\ &= 3a^2(1+\cos\theta), \\ &= 6a^2\cos^2\frac{\theta}{2}. \end{aligned}$$

But
$$\varrho = \frac{(x'^2+y'^2)^{3/2}}{x'y''-x''y'},$$
$$= \frac{4}{3}a\cos\frac{\theta}{2}.$$

Hence the radius of curvature is minimum for $\theta=\pi$,

and
$$\varrho_{\min} = 0.$$

The radius of curvature is maximum for $\theta=0$,

and
$$\varrho_{\max} = \frac{4}{3}a.$$

EXERCISE 33

1. Find the radius of curvature of the curve $y = x^3 - 2x$ at the point $A(1, -1)$.

2. Find the radius of curvature of the parabola $y^2 = px$, $p > 0$, at the origin.

3. Find the radius of curvature of the parabola $y = x^2$ at the point $x = 2$.

4. Find the radius of curvature of the parabola $y^2 = 4x\sqrt{2}$ at the point $x = \sqrt{2}$.

5. Find the curvature of the semi-cubical parabola $y^2 = x^3$ at the point $x = 4/3$.

6. Find the curvature of the ellipse $x^2/a^2 + y^2/b^2 = 1$, $a > 0$, $b > 0$, at the points $x = 0$ and $x = a$.

7. Find the points of the hyperbola $xy = 1$ at which the curvature is greatest.

Find the radii of curvature of the following curves at the points indicated:

8. $3a^2y = x^3$, for $y = 1$.

9. $y = x^4 - 4x^3 - 18x^2$, for $x = 0$.

10. The hyperbola $x^2/9 - y^2/4 = 1$, for $x = 9$.

11. $y^2(x-4m) = mx(x-3m)$, at the points of intersection of the curve with the x-axis.

12. $y = \tan x$, at the point $x = \pi/4$.

13. $y = a \log \cos \dfrac{x}{a}$, at a general point.

14. $y = 3 \sin 2x$, at the point $x = \pi/4$.

15. $y = 4 \cos \dfrac{x}{2}$, at the point $x = \pi/2$.

16. $y = 2e^{3x}$, at the point $x = 0$.

17. $y = xe^{-x}$, at the point $x = -1$.

18. The catenary $y = c \cosh \dfrac{x}{c}$, at its lowest point.

19. $y = \sin x$, at the point $x = \pi/2$.

20. $x = a(\cos t + t \sin t)$, $y = a(\sin t - t \cos t)$, $a > 0$, at the point $t = t_0$.

21. $x = t^2$, $y = t^3$, at the point $(1, 1)$.

22. $x = a(t - \sin t)$, $y = a(1 - \cos t)$, at the point $t = \pi/3$.

23. Find the extreme values of the radius of curvature of the curve $r = a \cos^3 \dfrac{\theta}{3}$, $a > 0$, in the interval $0 \leqslant \theta \leqslant 3\pi$. [*Hint:* transform to cartesian coordinates.]

24. $r = 2a \cos \theta$, $a > 0$, at the point $\theta = \theta_0$. [*Hint:* transform to cartesian coordinates.]

25. The Archimedean spiral $r = a\theta$, $a > 0$, at the point $\theta = \theta_0$.

26. $r^2 = a^2 \cos 2\theta$, for $\theta > 0$.

27. A particle of mass m slides down the smooth curve $y = x^2/2$. At the point $A(1, 1/2)$ the velocity of the particle is v. Find the force exerted by the particle on the curve (Fig. 89). (The required force N is normal to the curve and is given by $N = (mv^2)/\varrho$, where ϱ is the radius of curvature.)

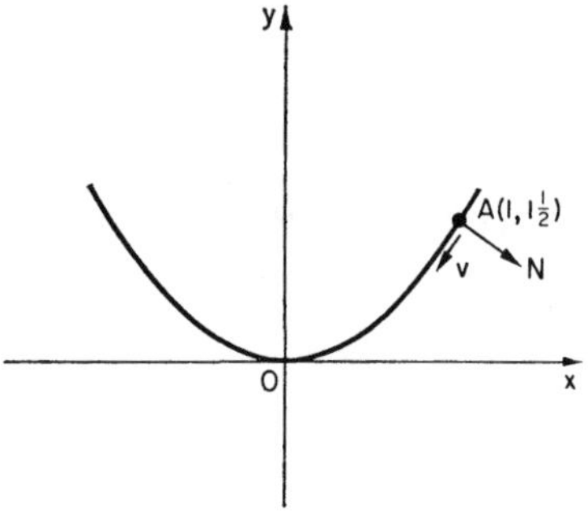

FIG. 89

28. The path of a moving particle is the ellipse $x = a \cos t$, $y = b \sin t$, where t is time, a and b are constants and $a \neq b$. Find the acceleration p perpendicular to the motion at time $t = 0$, where $p = v^2/\varrho$, v is the linear velocity and ϱ is the radius of curvature.

§ 3. Evolute and involute

3.1 Coordinates of the centre of curvature

(i) If the curve is given by the equation $y = f(x)$, then, assuming that $y'' \neq 0$, the centre of curvature at the point (x, y) of the curve has coordinates (ξ, η) where

$$\xi = x - y' \frac{1+y'^2}{y''},$$

$$\eta = y + \frac{1+y'^2}{y''}.$$

(ii) If the curve is given in parametric form $x = g(t)$, $y = h(t)$, then, assuming that $\dot{x}\ddot{y} - \ddot{x}\dot{y} \neq 0$, the centre of curvature at the point t has coordinates (ξ, η) where

$$\xi = x - \dot{y}\,\frac{\dot{x}^2+\dot{y}^2}{\dot{x}\ddot{y}-\ddot{x}\dot{y}},$$

$$\eta = y + \dot{x}\,\frac{\dot{x}^2+\dot{y}^2}{\dot{x}\ddot{y}-\ddot{x}\dot{y}}.$$

3.2 Definition

If the radius of curvature of the curve L changes continuously then the locus of the centre of curvature of L is another curve L'. The second curve L' is the *evolute* of the curve L. The first curve L is an *involute* of the curve L'. Note that a curve has only one evolute, whereas it has infinitely many involutes. Referring to Fig. 90 we see that any curve "parallel" to L (in the sense that tangents at similar points are parallel) is an involute of L'.

To find the evolute of $y = f(x)$ we express y, y', y'' in terms of x, substitute in the two formulae for ξ and η, and by eliminating x between these two equations obtain the equation of the evolute (the locus of (ξ, η)) in the form $F(\xi, \eta) = 0$.

If the equation is given in parametric form we eliminate the parameter t between the two equations and thus obtain the required equation.

3.3 Two properties of the evolute

(i) The normal to the given curve is tangent to the evolute.

(ii) Given a curve L and its evolute L', such that to points P and Q on L correspond points P' and Q' on L', the arc length $P'Q'$ is equal to the numerical difference between the radii of curvature at P and Q, i.e. (Fig. 90),

$$\text{arc } P'Q' = |PP' - QQ'|.$$

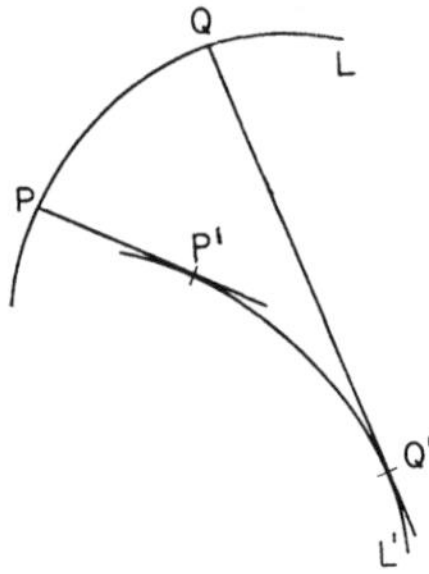

FIG. 90

Examples

1. Find the equation of the circle of curvature of the cissoid

$$y^2 = \frac{x^3}{2a-x}, \quad a > 0, \text{ at the point } x = \frac{2a}{3}.$$

Solution. We assume that $0 < x < 2a$.

Write the equation in the form

$$(2a-x)y^2 = x^3. \tag{i}$$

Differentiate

$$(2a-x)2yy' - y^2 = 3x^2. \tag{ii}$$

Differentiate again

$$(2a-x)2yy'' + (2a-x)2y'^2 - 2yy' - 2yy' = 6x,$$

or

$$(2a-x)yy'' + (2a-x)y'^2 - 2yy' = 3x. \tag{iii}$$

Substituting $x = 2a/3$ in equation (i) we obtain

$$y = \frac{a\sqrt{2}}{3}, \quad \text{or} \quad y = -\frac{a\sqrt{2}}{3}.$$

From (ii) we obtain the corresponding values

$$y' = \frac{7}{4\sqrt{2}}, \quad \text{or} \quad y' = -\frac{7}{4\sqrt{2}},$$

and from (iii)

$$y'' = \frac{81}{32a\sqrt{2}}, \quad \text{or} \quad y'' = -\frac{81}{32a\sqrt{2}}.$$

But

$$\varrho = \frac{(1+y'^2)^{3/2}}{y''},$$

hence the radius of curvature at $x = 2a/3$ is

$$\varrho = \pm\frac{9a}{4}.$$

The coordinates of the centre of curvature are

$$\xi = x - y'\,\frac{1+y'^2}{y''}, \qquad \eta = y + \frac{1+y'^2}{y''},$$

therefore,

$$\xi = -\frac{13a}{12}, \quad \text{and} \quad \eta = \pm\frac{4a\sqrt{2}}{3},$$

and the two circles of curvature have equations

$$\left(x+\frac{13a}{12}\right)^2 + \left(y\pm\frac{4a\sqrt{2}}{3}\right)^2 = \frac{81a^2}{16}.$$

2. Find the circle of curvature of the parabola $y = x^2/2a - a$, $a > 0$, at the point $x = a$, and then write the equation of the evolute of this curve.

Solution. We find the first and second derivatives.

$$y' = \frac{x}{a}, \qquad y'' = \frac{1}{a}.$$

Therefore the radius of curvature is

$$\varrho = \frac{(1+y'^2)^{3/2}}{y''}$$

$$= \frac{(a^2+x^2)^{3/2}}{a^2},$$

and the coordinates of the centre of curvature are

$$\xi = x - y'\,\frac{1+y'^2}{y''} = -\frac{x^3}{a^2},$$

$$\eta = y + \frac{1+y'^2}{y''} = \frac{3x^2}{2a}.$$

Hence for $x = a$, $\varrho = 2a\sqrt{2}$, $\xi = -a$, $\eta = 3a/2$ and the equation of the circle of curvature is

$$(x+a)^2 + \left(y - \frac{3}{2}\,a\right)^2 = 8a^2.$$

Eliminating x between the two expressions for ξ and η we obtain the relationship

$$\eta^3 = \frac{27}{8}\,a\xi^2,$$

hence the equation of the evolute of the given parabola is

$$y^3 = \frac{27}{8}\,ax^2,$$

which is a semi-cubical parabola.

3. Find the evolute of the astroid

$$x^{2/3} + y^{2/3} = a^{2/3}.$$

Solution. The parametric equations of the astroid are

$$x = a\cos^3 t, \qquad y = a\sin^3 t.$$

Thus,

$$\dot{x} = -3a\sin t\cos^2 t, \qquad \ddot{x} = 6a\sin^2 t\cos t - 3a\cos^3 t.$$

$$\dot{y} = 3a\sin^2 t\cos t, \qquad \ddot{y} = 6a\sin t\cos^2 t - 3a\sin^3 t.$$

Thus,

$$\dot{x}^2 + \dot{y}^2 = 9a^2\sin^2 t\cos^2 t,$$

and

$$\dot{x}\ddot{y} - \ddot{x}\dot{y} = -9a^2\sin^2 t\cos^2 t.$$

Thus, if (ξ, η) are the coordinates of the centre of curvature,

$$\xi = a\cos^3 t - \frac{3a\sin^2 t\cos t\,.\,9a^2\sin^2 t\cos^2 t}{-9a^2\sin^2 t\cos^2 t},$$

$$= a\cos^3 t + 3a\sin^2 t\cos t,$$

similarly,

$$\eta = a\sin^3 t + 3a\sin t\cos^2 t.$$

Thus,

$$\xi + \eta = a(\cos t + \sin t)^3,$$
$$\xi - \eta = a(\cos t - \sin t)^3,$$

therefore,

$$(\xi+\eta)^{2/3} + (\xi-\eta)^{2/3} = 2a^{2/3},$$

and the equation of the evolute is

$$(x+y)^{2/3} + (x-y)^{2/3} = 2a^{2/3}.$$

Rotate the axes through $\pi/4$ so that referred to the original axes the equations of the new axes are $x+y=0$ and $x-y=0$ Put $x+y = X$, and $x-y = -Y$, then the equation of the evolute of the astroid

$$x^{2/3} + y^{2/3} = a^{2/3},$$

is the astroid $$X^{2/3} + Y^{2/3} = 2a^{2/3}.$$

4. Find the evolute of the curve whose parametric form is $x = a(\cos t + t\sin t)$, $y = a(\sin t - t\cos t)$, $a > 0$.

Solution. From the given relations

$$\dot{x} = at\cos t, \qquad \ddot{x} = a(\cos t - t\sin t),$$
$$\dot{y} = at\sin t, \qquad \ddot{y} = a(\sin t + t\cos t),$$
$$\dot{x}^2 + \dot{y}^2 = a^2t^2, \qquad \dot{x}\ddot{y} - \ddot{x}\dot{y} = a^2t^2.$$

Hence, $$\xi = x - \dot{y}\,\frac{\dot{x}^2+\dot{y}^2}{\dot{x}\ddot{y}-\ddot{x}\dot{y}} = a\cos t,$$

$$\eta = y + \dot{x}\,\frac{\dot{x}^2+\dot{y}^2}{\dot{x}\ddot{y}-\ddot{x}\dot{y}} = a\sin t.$$

Eliminating the parameter t we obtain the relationship

$$\xi^2+\eta^2 = a^2,$$

hence the equation of the evolute is

$$x^2+y^2 = a^2.$$

Thus the given curve is an involute of a circle.

5. Find the evolute of the logarithmic spiral $r = ae^\theta$, $a > 0$.

Solution. Transforming into cartesian coordinates we obtain,

$$x = ae^\theta \cos\theta, \qquad y = ae^\theta \sin\theta.$$

Thus,

$$x' = ae^\theta (\cos\theta-\sin\theta), \qquad x'' = -2ae^\theta \sin\theta,$$
$$y' = ae^\theta (\sin\theta+\cos\theta), \qquad y'' = 2ae^\theta \cos\theta,$$
$$x'^2+y'^2 = 2a^2e^{2\theta}, \qquad x'y''-x''y' = 2a^2e^{2\theta}.$$

Therefore,

$$\xi = -\, ae^\theta \sin\theta, \quad \text{and} \quad \eta = ae^\theta \cos\theta.$$

Therefore the evolute of the logarithmic spiral

$$x = ae^\theta \cos\theta, \qquad y = ae^\theta \sin\theta,$$

is the curve

$$\xi = -ae^\theta \sin\theta, \qquad \eta = ae^\theta \cos\theta.$$

Rotating the axes through $\pi/2$ and writing

$$X = \xi, \qquad Y = -\eta,$$

we obtain the equation of the evolute in the form

$$X = ae^\theta \cos\theta, \qquad Y = ae^\theta \sin\theta.$$

It follows that the evolute of a logarithmic spiral is the same spiral rotated through an angle of $-\pi/2$.

EXERCISE 34

1. Find the equation of the circle of curvature for the parabola $y^2 = 12x$, at the point $P(3, -6)$.

2. Find the equations of the circles of curvature for the curves $y = \cos x$, and $y = \sec x$, at the point $x = 0$.

3. Find the equation of the circle of curvature of the cardioid $x = a \cos \theta(1+\cos \theta)$, $y = a \sin \theta(1+\cos \theta)$, $a > 0$, at the point $\theta = \pi/2$.

Find the evolutes of the following curves:

4. The parabola $y^2 = 4ax$, $a > 0$.

5. The ellipse $x^2/a^2+y^2/b^2 = 1$, $a > 0$, $b > 0$.

6. The hyperbola $x^2/a^2-y^2/b^2 = 1$, $a > 0$, $b > 0$.

7. The hyperbola $xy = c^2$.

8. $x = 3t^2$, $y = 3t-t^3$.

9. The catenary $y = c \cosh x/c$.

10. $y = a \log \cos x/a$.

11. The cycloid $x = a(t-\sin t)$, $y = a(1-\cos t)$, $a > 0$.

12. The tractrix $x = a(\cos t+\log \tan t/2)$, $y = a \sin t$, $a > 0$.

13. The hypocycloid $x = a(2 \cos t+\cos 2t)$, $y = a(2 \sin t-\sin 2t)$.

CHAPTER 14

FOURIER SERIES

§ 1. General ideas

1.1 The Euler—Fourier formulae

We are concerned here with series of the form

$$a_0+(a_1 \cos x+b_1 \sin x)+(a_2 \cos 2x+b_2 \sin 2x)+\dots$$
$$+(a_n \cos nx+b_n \sin nx)+\dots$$
$$=\sum_{n=0}^{\infty}(a_n \cos nx+b_n \sin nx), \tag{1}$$

where the coefficients a_n and b_n are constants.

Assuming that this series is uniformly convergent (see Chapter 7, 2.4) in the interval $-\pi \leqslant x \leqslant \pi$ we denote the sum of the series as $f(x)$. Then all the coefficients of this series can be expressed as integrals of $f(x)$ as follows:

$$a_0=\frac{1}{2\pi}\int_{-\pi}^{\pi} f(x)\,\mathrm{d}x,$$

$$\left.\begin{aligned} a_n&=\frac{1}{\pi}\int_{-\pi}^{\pi} f(x)\cos nx\,\mathrm{d}x\\ b_n&=\frac{1}{\pi}\int_{-\pi}^{\pi} f(x)\sin nx\,\mathrm{d}x\end{aligned}\right\}\quad \text{for } n=1, 2, \dots. \tag{2}$$

These are the Euler–Fourier formulae.

1.2 The Fourier series

(i) If $f(x)$ is a given integrable function in the interval $-\pi \leqslant \leqslant x \leqslant \pi$, and if the coefficients $a_0, a_1, \ldots, a_n$, and $b_1, b_2, \ldots, b_n$ are found from the given formulae (2) and substituted in (1), we obtain a series which we call the Fourier series of the function $f(x)$. We should note that a Fourier series so obtained is not necessarily convergent, or if convergent is not necessarily convergent to $f(x)$. We can also note that the coefficient a_0 is the mean value of the function $f(x)$ in the interval $[-\pi, \pi]$.

(ii) The Fourier series (1) can also be expressed in the form

$$A_0+A_1 \sin (x+\phi_1)+A_2 \sin (2x+\phi_2)+\ldots+ \\ +A_n \sin (nx+\phi_n)+\ldots \\ = A_0+\sum_{n=1}^{\infty} A_n \sin (nx+\phi_n). \qquad (3)$$

The relationships between the coefficients a_n, b_n of (1) and A_n, ϕ_n of (3) are as follows:

$$a_0 = A_0, \quad a_n = A_n \sin \phi_n, \quad b_n = A_n \cos \phi_n, \quad n = 1, 2, 3, \ldots,$$

and

$$A_n = \sqrt{(a_n^2+b_n^2)}, \qquad \phi_n = \text{arc tan } (a_n/b_n), \qquad n = 1, 2, 3, \ldots.$$

1.3 Dirichlet's conditions

If in the closed interval $[-\pi, \pi]$ the function $f(x)$ fulfills the following two conditions then in this interval the function is the sum of the corresponding Fourier series.

(i) The closed interval $[-\pi, \pi]$ can be divided into a finite number of sub-intervals such that at each point of these intervals $f(x)$ is monotonic.

(ii) At every point inside the open interval $(-\pi, \pi)$ the value of the function $f(x)$ is the mean value of the limit of the function from the left and the limit of the function from the right at this point, i.e.

$$f(x) = \frac{1}{2}\{f(x)^{-} + f(x)^{+}\}, \qquad \text{for } -\pi < x < \pi.$$

At each of the extremities of the interval ($x = -\pi$, $x = \pi$), the value of the function is equal to the arithmetic mean of the limit from the right at $x = -\pi$ and the limit from the left at $x = \pi$, i.e. $\frac{1}{2}\{f(-\pi)^{+} + f(\pi)^{-}\}$.

1.4 Definition

A function $f(x)$ defined in the closed interval $[a, b]$ is called regular in its sub-intervals if the given interval can be divided into a finite number of closed sub-intervals such that in each of these intervals the first derivative of the function is a continuous function.

If the function $f(x)$ is regular in its sub-intervals then the first of Dirichlet's conditions is automatically fulfilled.

If the function $f(x)$ is continuous inside the interval $[-\pi, \pi]$ then the second of Dirichlet's conditions is fulfilled since for every point of the interval $f(x)^{-} = f(x)^{+} = f(x)$.

1.5 Odd and even functions

If $f(x)$ is an even function, i.e. if $f(-x) = f(x)$, then in the expansion of the function as a Fourier series all the coefficients b_n are zero.

$$b_n = \frac{1}{\pi}\int_{-\pi}^{\pi} f(x) \sin nx \,\mathrm{d}x,$$

$$= \frac{1}{\pi}\int_{-\pi}^{0} f(x) \sin nx \,\mathrm{d}x + \frac{1}{\pi}\int_{0}^{\pi} f(x) \sin nx \,\mathrm{d}x,$$

$$= \frac{1}{\pi} \int_{\pi}^{0} f(-x) \sin(-nx)(-dx) + \frac{1}{\pi} \int_{0}^{\pi} f(x) \sin nx \, dx,$$

$$= \frac{1}{\pi} \int_{\pi}^{0} f(x) \sin nx \, dx + \frac{1}{\pi} \int_{0}^{\pi} f(x) \sin nx \, dx,$$

$$= 0.$$

In general, a_n will not be zero (since $\cos nx$ is an even function) and the Fourier series will be a series of cosines.

Similarly, if $f(x)$ is an odd function, i.e. if $f(-x) = -f(x)$ the coefficients a_n will be zero and the Fourier series of the function will be a series of sines.

1.6 *Functions of period 2l*

If we want to find the Fourier series of $g(t)$ defined in the interval $[-l, l]$, then by substituting $t = lx/\pi$, we obtain the function

$$f(x) = g(lx/\pi),$$

where $f(x)$ is defined in the interval $[-\pi, \pi]$. Then the expansion of $g(t)$ is

$$a_0 + \sum_{n=1}^{\infty} \left\{ a_n \cos \frac{\pi n t}{l} + b_n \sin \frac{\pi n t}{l} \right\}$$

where the coefficients can be calculated by the formulae:

$$a_0 = \frac{1}{2l} \int_{-l}^{l} g(t) \, dt,$$

$$\left. \begin{aligned} a_n &= \frac{1}{l} \int_{-l}^{l} g(t) \cos \frac{\pi n t}{l} \, dt, \\ b_n &= \frac{1}{l} \int_{-l}^{l} g(t) \sin \frac{\pi n t}{l} \, dt, \end{aligned} \right\} \text{ for } n > 1.$$

1.7 Functions defined in the interval [a, b]

It may happen that we want to find the Fourier series of the continuous function $f(x)$, regular in its sub-intervals, defined in the interval $[a, b]$ included in the interval $[-\pi, \pi]$. In this case we extend the function $f(x)$ over the whole interval. We define a function $g(x)$ in the interval $-\pi \leqslant x \leqslant \pi$, such that $g(x) \equiv f(x)$ for $a \leqslant x \leqslant b$. Such extension can be made in different ways but we choose such a way that the function $g(x)$ is also regular in its sub-intervals, e.g.

$$g(x) = \begin{cases} f(a), & \text{for} \quad -\pi \leqslant x \leqslant a, \\ f(x), & \text{for} \quad a \leqslant x \leqslant b, \\ f(b), & \text{for} \quad b \leqslant x \leqslant \pi. \end{cases}$$

In particular, when $f(x)$ is defined in the half-interval $0 \leqslant x \leqslant \pi$, we extend the function over the interval $[-\pi, \pi]$ in such a way that the extended function is either odd or even. The Fourier series will then be simplified.

If $f(0) = 0$ we may extend the function as an odd function by defining $f(x) = -f(-x)$ for $x < 0$. Then the coefficients a_n are zero, and

$$b_n = \frac{1}{\pi}\int_{-\pi}^{\pi} f(x) \sin nx \, dx,$$

$$= \frac{2}{\pi}\int_{0}^{\pi} f(x) \sin nx \, dx, \qquad n > 1.$$

Alternatively we may extend the function as an even function by defining $f(x) = f(-x)$ for $x < 0$. Then the coefficients b_n are zero, and

$$a_0 = \frac{1}{\pi}\int_{0}^{\pi} f(x) \, dx,$$

$$a_n = \frac{2}{\pi}\int_{0}^{\pi} f(x) \cos nx \, dx, \qquad n > 1.$$

Examples

1. Find the Fourier series for $f(x)=|x|$ in the interval $[-\pi, \pi]$

Solution. From the graph of the function (Fig. 91) we can see that the first of Dirichlet's conditions is fulfilled since the interval $[-\pi, \pi]$ can be divided into two sub-intervals such that the function is monotonic decreasing in the one $(-\pi \leqslant x \leqslant 0)$, and monotonic increasing in the other $(0 \leqslant x \leqslant \pi)$.

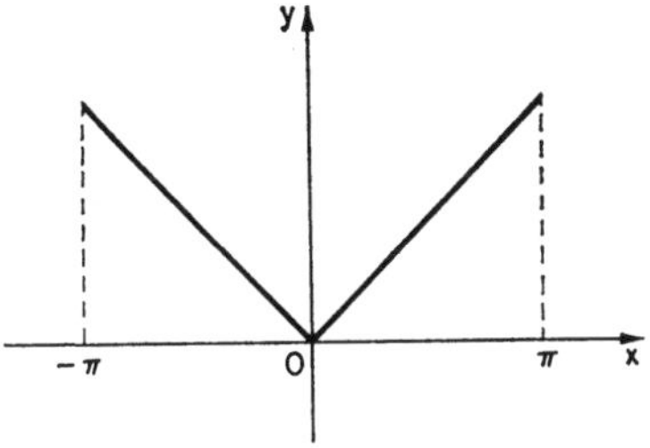

FIG. 91

For the second condition it is enough to note that this function is continuous for all values of x. Also as $f(-\pi)=f(\pi)=\pi$ the second Dirichlet condition is fulfilled at the extremities of the interval. Thus,

$$a_0 = \frac{1}{2\pi}\int_{-\pi}^{\pi} |x|\,dx,$$

$$= \frac{1}{2\pi}\int_{-\pi}^{0} (-x)dx + \frac{1}{2\pi}\int_{0}^{\pi} x\,dx,$$

$$= \frac{1}{2}\pi.$$

$$a_n = \frac{1}{\pi}\int_{-\pi}^{\pi} |x| \cos nx\,dx,$$

$$= \frac{1}{\pi}\int_{-\pi}^{0} (-x)\cos nx\,dx + \frac{1}{\pi}\int_{0}^{\pi} x \cos nx\,dx,$$

$$= \frac{2}{\pi}\int_{0}^{\pi} x \cos nx\,dx.$$

Integrating by parts we obtain

$$a_n = \frac{2}{\pi}\left[\frac{x}{n}\sin nx + \frac{1}{n^2}\cos nx\right]_0^\pi,$$

$$= \frac{2}{\pi}\left(\frac{1}{n^2}\cos n\pi - \frac{1}{n^2}\right),$$

therefore,

$$a_n = \begin{cases} 0, & \text{for } n \text{ even} \\ -\dfrac{4}{\pi n^2}, & \text{for } n \text{ odd.} \end{cases}$$

$f(x) = |x|$ is an even function of x, hence the coefficients b_n are zero. Therefore,

$$|x| = \frac{1}{2}\pi - \frac{4\cos x}{\pi . 1^2} - \frac{4\cos 3x}{\pi . 3^2} - \frac{4\cos 5x}{\pi . 5^2} - \ldots,$$

$$= \frac{1}{2}\pi - \frac{4}{\pi}\sum_{n=1}^{\infty}\frac{\cos(2n-1)x}{(2n-1)^2}.$$

The series is valid for all values of x lying in the interval $-\pi \leqslant x \leqslant \pi$. For $x = 0$ we obtain

$$0 = \frac{1}{2}\pi - \frac{4}{\pi}\left(\frac{1}{1^2} + \frac{1}{3^2} + \frac{1}{5^2} + \ldots\right),$$

which gives us

$$\frac{\pi^2}{8} = \frac{1}{1^2} + \frac{1}{3^2} + \frac{1}{5^2} + \ldots + \frac{1}{(2n-1)^2} + \ldots.$$

Outside the given interval $[-\pi, \pi]$ the sum of the series is no longer equal to $|x|$ because the right hand side is periodic with period 2π, whereas $|x|$ is not periodic. The graph of the periodic function is shown in Fig. 92.

2. Express as a Fourier series the function

$$f(x) = \begin{cases} 0, & \text{for } -\pi < x \leqslant 0 \\ x, & \text{for } 0 \leqslant x < \pi. \end{cases}$$

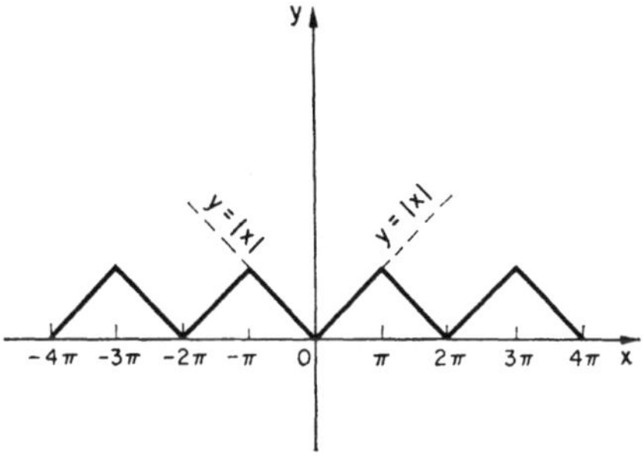

FIG. 92

Solution. The graph of the periodic function obtained by extending the given function outside the range $[-\pi, \pi]$ is shown in Fig. 93. Dividing the interval into the two appropriate intervals $[-\pi, 0]$ and $[0, \pi]$, we obtain

$$a_0 = \frac{1}{2\pi}\int_{-\pi}^{0} 0\,dx + \frac{1}{2\pi}\int_{0}^{\pi} x\,dx,$$

$$= \frac{1}{4}\pi.$$

$$a_n = \frac{1}{\pi}\int_{-\pi}^{0} 0\,dx + \frac{1}{\pi}\int_{0}^{\pi} x\cos nx\,dx,$$

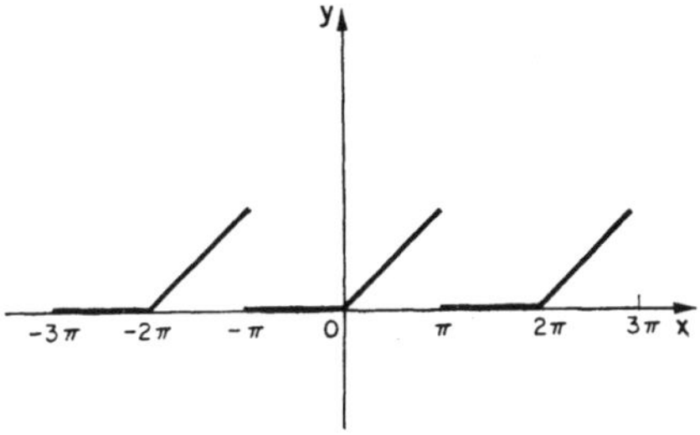

FIG. 93

and integrating by parts we obtain

$$a_n = \begin{cases} 0, & \text{for } n \text{ even} \\ -\dfrac{2}{\pi n^2}, & \text{for } n \text{ odd.} \end{cases}$$

Similarly,

$$b_n = \frac{1}{\pi}\int_{-\pi}^{0} 0\,dx + \frac{1}{\pi}\int_{0}^{\pi} x \sin nx\,dx,$$

$$= \frac{1}{\pi}\left[-\frac{x}{n}\cos nx + \frac{1}{n^2}\sin nx\right]_0^{\pi},$$

thus,

$$b_n = \begin{cases} -\dfrac{1}{n}, & \text{for } n \text{ even} \\ +\dfrac{1}{n}, & \text{for } n \text{ odd.} \end{cases}$$

Therefore,

$$f(x) = \frac{1}{4}\pi + \left(-\frac{2}{\pi}\cos x + \sin x\right) + \left(-\frac{\sin 2x}{2}\right)$$

$$+\left(-\frac{2}{9\pi}\cos 3x + \frac{\sin 3x}{3}\right) + \left(-\frac{\sin 4x}{4}\right) + \ldots$$

$$+\left(-\frac{2}{(2n-1)^2\,\pi}\cos(2n-1)x + \frac{\sin(2n-1)x}{2n-1}\right)$$

$$+\left(-\frac{\sin 2nx}{2n}\right) + \ldots.$$

At $x = \pi$ the value of the function is not convergent to π, but according to Dirichlet's second condition

$$f(\pi) = \frac{1}{2}(0+\pi),$$

$$= \frac{1}{2}\pi.$$

But $$f(\pi) = \frac{1}{4}\pi + \frac{2}{\pi} + \frac{2}{9\pi} + \ldots + \frac{2}{(2n-1)^2\pi} + \ldots,$$

thus, $$\frac{1}{2}\pi = \frac{1}{4}\pi + \frac{2}{\pi}\left\{\frac{1}{1^2} + \frac{1}{3^2} + \frac{1}{5^2} + \ldots + \frac{1}{(2n-1)^2} + \ldots\right\},$$

or $$\frac{\pi^2}{8} = \frac{1}{1^2} + \frac{1}{3^2} + \frac{1}{5^2} + \ldots + \frac{1}{(2n-1)^2} + \ldots,$$

as above.

It can be shown that the series is absolutely convergent and consequently we may collect the terms and write

$$f(x) = \frac{1}{4}\pi - \frac{2}{\pi}\sum_{n=1}^{\infty}\frac{\cos(2n-1)x}{(2n-1)^2} + \sum_{n=1}^{\infty}(-1)^{n+1}\frac{\sin nx}{n}.$$

3. Find the amplitudes of the harmonic components of a set of oscilloscope time base potentials. The graph of the potential variation $U(t)$ is a saw-tooth curve, where for $-\pi < t < \pi$ the gradient of the curve is 1.

Solution. Since the gradient of the curve is 1, the function is of the form

$$U(t) = t.$$

The graph of the function is shown as a periodic function of period 2π (Fig. 94). Since the function is an odd function the

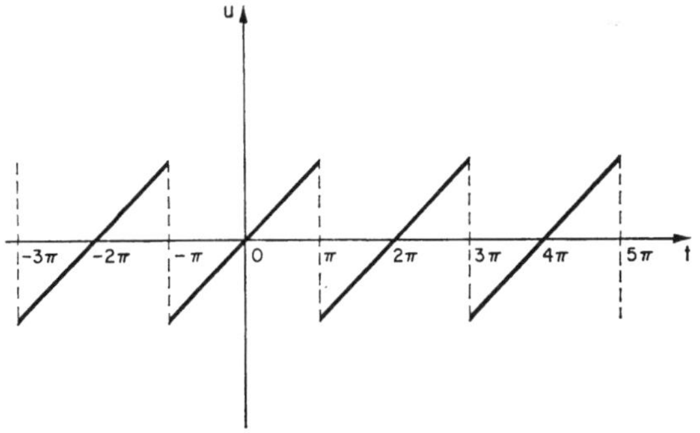

FIG. 94

coefficients a_n are zero for all n.

$$\begin{aligned} b_n &= \frac{1}{\pi}\int_{-\pi}^{\pi} t \sin nt \, dt, \\ &= \frac{1}{\pi n^2}\Big[-nt\cos nt + \sin nt\Big]_{-\pi}^{\pi}, \\ &= -\frac{2}{n^2}\cos n\pi, \\ &= (-1)^{n+1}\frac{2}{n}. \end{aligned}$$

The coefficient b_n is the amplitude of the n harmonic components of the corresponding potential. The expansion of the potential function as a Fourier series is

$$U(t) = 2\left(\sin t - \frac{1}{2}\sin 2t + \ldots + (-1)^{n+1}\frac{1}{n}\sin nt + \ldots\right).$$

EXERCISE 35

1. Express as a Fourier series the function $f(x) = x$ in the interval $[-\pi, \pi]$. What value does the series give for $x = -\pi$ and $x = \pi$?

2. Expand as a Fourier series (as a series of sines) $f(x) = a$, in the interval $0 < x < \pi$.

Express as Fourier series the functions:

3. $$f(x) = \begin{cases} 0, & 0 < x < \alpha, \quad \pi-\alpha < x < \pi \\ a, & \alpha < x < \pi-\alpha. \end{cases}$$

4. $$f(x) = \begin{cases} \dfrac{ax}{\alpha}, & 0 \leqslant x \leqslant \alpha, \\ a, & \alpha \leqslant x \leqslant \pi-\alpha, \\ \dfrac{a(\pi-x)}{\alpha}, & \pi-\alpha \leqslant x \leqslant \pi. \end{cases}$$

5. $$f(x) = \begin{cases} -c, & -\pi < x < 0, \\ +c, & 0 < x < \pi, \end{cases} \qquad \text{where } c \text{ is positive.}$$

6. $$f(x) = \begin{cases} 2x, & -\pi < x < 0, \\ 0, & x = 0, \\ 6x, & 0 < x < \pi. \end{cases}$$

What is the value of this series at the points $x = -\pi$ and $x = \pi$?

7. Show that in the interval $-\pi/2 < x < \pi/2$,

$$\frac{\pi x}{4} = \frac{\sin x}{1^2} - \frac{\sin 3x}{3^2} + \frac{\sin 5x}{5^2} - \ldots .$$

8. Express both as a series of sines and as a series of cosines the function $f(x) = x(\pi - x)$ in the interval $0 < x < \pi$.

9. Express as Fourier series the function $f(x) = |\sin x|$ for any value of x.

10. Express as a Fourier series the function $f(x) = x^2$, for $-\pi \leqslant x \leqslant \pi$.

11. Express as a Fourier series the function $f(x) = x^3$, for $-\pi \leqslant x \leqslant \pi$.

12. Express as a Fourier series the function $f(x) = \sin ax$, for $-\pi < x < \pi$, where a is not an integer.

13. Express as a Fourier series the function

$$f(x) = \begin{cases} 0, & -\pi \leqslant x \leqslant 0, \\ \sin x, & 0 \leqslant x \leqslant \pi. \end{cases}$$

14. Express as a Fourier series the function

$$f(x) = \begin{cases} -\cos x, & -\pi < x < 0, \\ 0, & x = 0, \\ \cos x, & 0 < x < \pi. \end{cases}$$

15. Express as a Fourier series the function $f(x) = x \cos x$ in the interval $0 < x < \pi$.

16. Express as a Fourier series (as a series of cosines) the function $f(x) = (1/2) \log \cot x/2$ in the interval $0 < x < \pi$.

17. Express as a series of cosines the function, $f(x) = \log (2 \cos x/2)$ in the interval $0 \leqslant x < \pi$.

18. Express as a series of cosines the function, $f(x) = -\log (2 \sin x/2)$ in the interval $0 < x \leqslant \pi$.

19. Express as a Fourier series the function $f(x) = \cos ax$ in the interval $-\pi < x < \pi$, where a is any non-integral number.

20. Express as a series of cosines $f(x) = \cos x$, in the interval $0 < x < \pi$.

21. A single diode valve rectifier supplies direct current according to the formula

$$i = \begin{cases} I \sin nt, & \text{for } 0 \leqslant t \leqslant \pi/n, \\ 0, & \text{for } -\pi/n \leqslant t \leqslant 0, \end{cases}$$

where n is the frequency of oscillation, t is time and I is a constant. Calculate the amplitudes of the harmonic components of the rectified current.

22. Calculate the amplitudes of the harmonic components of the magnetic induction due to a rotating magnetic field $B = f(x)$, given that the form of the induction curve is as in Fig. 95.

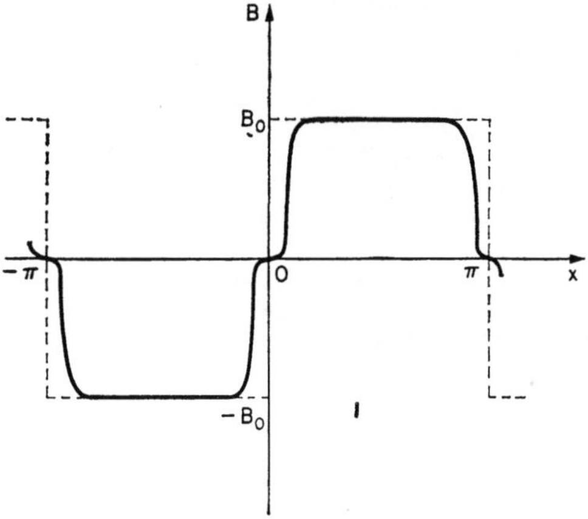

FIG. 95

In order to express this as a Fourier series the curve must first be "rectified" as shown by the dotted line. Then the function to be considered is

$$f(x) = \begin{cases} -B_0, & \text{for } -\pi < x < 0, \\ +B_0, & \text{for } 0 < x < \pi. \end{cases}$$

ANSWERS

EXERCISE 18, P. 11

1. $I = \frac{5}{3}x^3 - 3x^2 + 3x - 2\log|x| - \frac{5}{x} + C.$

2. $I = \frac{1}{6}x^6 - \frac{3}{4}x^4 + \frac{3}{2}x^2 - \log|x| + C.$

3. $I = \frac{1}{5}x^5 - \frac{1}{3}x^3 + x + C.$

4. $I = \frac{1}{12}(x^2+4)^6 + C.$

5. $I = \frac{1}{2}\log(1+x^2) + C.$

6. $I = \frac{-1}{10(x^2+3)^5} + C.$

7. $I = \frac{1}{3}\log(a^3+x^3) + C, \quad x \neq -a.$

8. $I = 3x^{1/3} - \frac{4}{3}x^{-3/4} + C, \quad x > 0.$

9. $I = \frac{6}{13}x^{13/6} - \frac{12}{23}x^{23/12} + C, \quad x > 0.$

10. $I = 27x + \frac{216}{5}x^{5/4} + 24x^{3/2} + \frac{32}{7}x^{7/4} + C.$

11. $I = \frac{1}{7}x^{7/6} - \frac{1}{4}x^{4/3} + \frac{8}{17}\sqrt[12]{(125x^{17})} + C.$

12. $I = -\frac{6}{\sqrt{x}} + 3\sqrt[6]{x} + C.$

13. $I = \frac{2}{9}(3x+1)^{3/2} + C, \quad x > \frac{1}{3}.$

14. $I = \frac{2}{36}\sqrt{(a+bx)^3} + C, \quad$ (put $a+bx = t^2$).

15. $I = \frac{3}{8}(2x^2-1)^{2/3} + C, \quad x \neq \frac{1}{\sqrt{2}}, \quad x \neq -\frac{1}{\sqrt{2}}.$

16. $I = \frac{1}{3}\sqrt{(1+x^2)^3} + C.$

17. $I = -\frac{\sqrt{(3-5x^2)}}{5} + C.$

18. $I = \frac{3}{5}(x-4)\sqrt[3]{(x+1)^2} + C.$

19. $I = \sqrt{(x^2-6)} + C.$

20. $I = \frac{5}{12}\sqrt[5]{(x^3+1)^4} + C.$

21. $I = -e^{1/x} + C.$

22. $I = -\frac{1}{2}e^{-x^2} + C.$

23. $I = \frac{1}{6}\tan 3x + C.$

24. $I = -\frac{1}{4}\cos(2x^2+1) + C.$

25. $I = \frac{1}{6}\sin^6 x + C.$

26. $I = 2\sqrt{(1+\sin x)} + C.$

27. $I = -\frac{1}{b}\log|a+b\cos x| + C, \quad a+b\cos x \neq 0.$

28. $I = e^{\sin x}+C.$

29. $I = \frac{1}{4}\tan(x^4)+C, \quad \cos(x^4) \neq 0.$

30. $I = \frac{1}{4}\tan^4 x+C, \quad \cos x \neq 0.$

31. $I = \frac{1}{3}\tan(x^3+1)+C, \quad \cos(x^3+1) \neq 0.$

32. $I = \frac{1}{3}(\log x)^3+C, \quad x > 0.$

33. $I = \arctan e^x+C, \quad (\text{put } e^x = t).$

34. $I = \frac{1}{2}\log(2e^x+1)+C.$

35. $I = \frac{1+x^2}{2}\log(1+x^2)-\frac{x^2}{2}+C.$

36. $I = \frac{2}{3}\sqrt{(2+\log|x|)^3}+C.$

37. $I = -\frac{6^{1-x}}{\log 6}+C.$

38. $I = \arcsin(\log|x|)+C.$

39. $I = \arctan x(\log|\arctan x|-1)+C.$

40. $I = \frac{1}{2}x^2e^{x^2}+C.$

41. $I = \frac{1}{3}\arcsin(x^3)+C, \quad -1 < x < 1.$

42. $I = \log|\arctan x|+C, \quad x = 0.$

43. $I = \pi \arcsin x-\frac{1}{2}(\arcsin x)^2+C, \quad -1 < x < 1.$

44. $I = \frac{1}{2}\arctan(x^2)+C.$

45. $I=\frac{1}{5}x^5(1+x)^3-\frac{1}{10}x^6(1+x)^2+\frac{1}{5}\left(\frac{1}{7}x^7+\frac{1}{8}x^8\right)+C.$

46. $I=(x^2-2x+2)\mathrm{e}^x+C.$

47. $I=(x^3-3x^2+6x-6)\mathrm{e}^x+C.$

48. $I=\mathrm{e}^{2x}\left(\frac{1}{2}x^4-x^3+\frac{3}{2}x^2-\frac{3}{2}x+\frac{3}{4}\right)+C.$

49. $I=x\sin x+\cos x+C.$

50. $I=(x^2-2)\sin x+2\cos x+C.$

51. $I=-\frac{1}{5}x^2\cos 5x+\frac{2}{25}x\sin 5x+\frac{2}{125}\cos 5x+C.$

52. $I=\frac{1}{2}\mathrm{e}^x(\sin x+\cos x)+C.$

53. $I=-\frac{1}{13}\mathrm{e}^{-2x}(2\sin 3x+3\cos 3x)+C.$

54. $I=-\frac{9}{13}\mathrm{e}^{-x}\left(-\cos\frac{2}{3}x+\frac{2}{3}\sin\frac{2}{3}x\right)+C.$

55. $I=\frac{2}{3}x^{3/2}\left(\log x-\frac{2}{3}\right)+C.$

56. $I=x\left\{(\log|x|)^3-3(\log|x|)^2+6\log|x|-6\right\}+C.$

57. $I=-\frac{1}{32x^4}(\log|x|^2+4\log|x|+1)+C.$

58. $I=\frac{2}{27}x^{3/2}\{9(\log|x|)^3-18(\log|x|)^2+24\log|x|-16\}+C.$

59. $I=-\frac{1}{3x^3}\left(\log|x|+\frac{1}{3}\right)+C.$

60. $I=2\sqrt{x}(\log x)^2-8\sqrt{x}\log x+16\sqrt{x}+C.$

61. $I=\frac{1}{4}x^4\left\{(\log x)^4-\frac{1}{2}\log x+\frac{1}{8}\right\}+C,\quad x>0.$

62. $I=\frac{x^n+1}{n+1}\left(\log x-\frac{1}{n+1}\right)+C,\quad x>0.$

EXERCISE 19, P. 29

1. $I=(2x+1)^4/8+C.$
2. $I=-1/9(3x-2)^3+C.$
3. $I=\log|x-3|+2\log|x+2|+C.$
4. $I=\log|x^2-3x+3|+C.$
5. $I=3\log|x-5|-2\log|x+1|+C.$
6. $I=5\log|x+1/2|-4\log|x+1|+C.$
7. $I=2\log|x-3|+4\log|x-1/2|+C.$
8. $I=\log|2x^2-5x+3|+C.$
9. $I=3\log|x-2|+2\log|x+5|+C.$
10. $I=(7\log|x-6|)/3-(3\log|x-3|)/2+C.$
11. $I=\{\log|(x+1-\sqrt{2})/(x+1+\sqrt{2})|\}/2\sqrt{2}+C.$
12. $I=\log|(2x-3)/(3x-2)|+C.$
13. $I=(\log|10x+x^2|)/2+C.$
14. $I=(7\log|4+5x^2|)/10+C.$
15. $I=\{\log|(x-1)/(5-x)|\}/4+C.$
16. $I=\{\log|(2x+\sqrt{5}-1)/(-2x+\sqrt{5}+1)|\}/\sqrt{5}+C.$
17. $I=\{\log|x/(2-3x)|\}/2+C.$
18. $I=(8\log|x-2|)/3+(\log|x+1|)/3+C.$
19. $I=2\log|x-3|-5/(x-3)+C.$
20. $I=\{\log|2x-1|+1/(2x-1)\}/4+C.$
21. $I=2\log|x-5|+3/(x-5)+C.$
22. $I=3\log|x+2|+5/(x+2)+C.$
23. $I=\{\arctan(2x-1)/3\}/3+C.$
24. $I=\{\arctan(3x+1)/\sqrt{2}\}/\sqrt{2}+C.$
25. $I=\{\arctan(x-3)/2\}/2+C.$
26. $I=\arctan(3x-1)+C.$
27. $I=(\log|x^2-x+1|)/2+\sqrt{3}\arctan(2x-1)/\sqrt{3}+C.$
28. $I=\log|2x^2-2x+1|+\arctan(2x-1)+C.$

29. $I = \log |x^2-2x+5| + \{\text{arc tan } (x-1)/2\}/2 + C.$

30. $I = \log |x^2+2x+10| - 4 \text{ arc tan } (x+1)/3 + C.$

31. $I = \log |x^2-8x+25| - 4 \text{ arc tan } (x-4)/3 + C.$

32. $I = (3 \log |x^2+4x+8|)/2 - \text{arc tan } (x+2)/2 + C.$

33. $I = (\log |x^2-3|)/2 + \sqrt{3} \log |(x-\sqrt{3})/(x+\sqrt{3})| + C.$

34. $I = (\log |x^2+3|)/2 + 2\sqrt{3} (\text{arc tan } x/\sqrt{3}) + C.$

35. $I = 3 \log |x^2+4x+13| - 4 \text{ arc tan } (x+2)/3 + C.$

36. $I = 5 \log |x^2-4x+20| - 6 \text{ arc tan } (x-2)/4 + C.$

37. $I = 2 \log |x^2-6x+10| + 7 \text{ arc tan } (x-3) + C.$

38. $I = 5x/3 - (10 \log |2+3x|)/9 + C.$

39. $I = x/5 - \{2\sqrt{15} \text{ arc tan } \sqrt{(5/12)}x\}/25 + C.$

40. $I = 2x - (5 \log |x^2+6x+25|)/2 - \{15 \text{ arc tan } (x+3)/4\}/4 + C.$

41. $I = -3 \log |x+2| + 2 \log |x-4| + 8 \log |x-7| + C.$

42. $I = -(29-30x+6x^2)/3(x-2)^3 + \log |x-2| + C.$

43. $I = (4 \log |x-2|)/7 + (17 \log |x^2+3|)/14 + (\text{arc tan } x/\sqrt{3})/7\sqrt{3} + C.$

44. $I = (\text{arc tan } x)/2 + (x-2)/2(x^2+1) + C.$

45. $I = x^2/2 + x + 3 \log |x+1| + 2 \log |x-2| + C.$

46. $I = x^2 - 3x - 14/(x-4) + 2 \log |x-4| + C.$

47. $I = x^3/3 - x + \text{arc tan } x + C.$

48. $I = 9x^5/5 - 2x^3 + 4x - \{8 \text{ arc tan } \sqrt{(3/2)}x\}/\sqrt{6} + C.$

49. $I = 2x^3/3 - 2x^2 + 5x + \log |x-2| + 2 \log |x-1| + C.$

50. $I = (5 \log |x+3|)/2 + (47 \log |x-1|)/2 - 25 \log |x - 1/2| + C.$

51. $I = (5 \log |x-1| - 4 \log |x+1| + 10 \log |x-2| - 11 \log |x+2|)/3 + C.$

52. $I = \{36x/(5-6x) - 12 \log |5/x-6| - 5/x + 6\}/25 + C.$

53. $I = \frac{1}{2} \log (x^2+1)/(x^2+3) + C.$

54. $I = 2\log(x^2-4x+29)+3\log(x^2-2x+5)$
$+6 \text{ arc tan } (x-2)/5+8 \text{ arc tan } (x-1)/2+C.$

55. $I = \log|x^2+4|+(5\log|x+1|)/2-\frac{1}{2}\log|x-1|$

$+\frac{1}{2} \text{ arc tan } \frac{1}{2}x+C.$

56. $I = \log|x|+2\log|x+1|+3\log|x+2|+4\log|x+3|+C.$

57. $I = 4\log|x|+\log(x^2+1)-1/x-\text{arc tan } x+C.$

58. $I = -(\log|x^2-a^2|)/2a^2-(\log|x|)/a^2+C.$

59. $I = \log|x|-\frac{1}{2}\log|x^2+x+1|-\{\text{arc tan } (2x+1)/\sqrt{3}\}$
$/\sqrt{3}+C.$

60. $I = \{\log(x^2+x+1)/(x^2-x+1)\}/4$
$+\{\text{arc tan } (2x+1)/\sqrt{3}\}/2\sqrt{3}+C.$

61. $I = 2\log|x^2-4|+\frac{1}{2}\log|x^2+4|+\left(3 \text{ arc tan } \frac{1}{2}x\right)/2+C.$

62. $I = 3\log|x-1|+6\log|x^2+4x+29|$
$+\{42 \text{ arc tan } (x+2)/5\}/5+C.$

63. $I = 1/(x+1)^2+\log|x|+3\log|x+1|+C.$

64. $I = 1/2x^2-1/x+2\log|x|-(\log|x+1|)/4+C.$

65. $I = (2x+1)/3(x^2+x+1)+4\{\text{arc tan } (2x+1)/\sqrt{3}\}/3\sqrt{3}+C.$

66. $I = 1/2(x-2)^2+5/(x-2)+3\log|x-2|+C.$

67. $I = (x+2)/16(x^2+4x+8)^2+3(x+2)/128(x^2+4x+8)$
$+\left\{3 \text{ arc tan } \frac{1}{2}(x+2)\right\}/256+C.$

68. $I = -5/2(x-1)-\log|x-1|+3/2(x+1)+2\log 2|x+1|$
$+C.$

69. $I = \left\{\frac{1}{2}\log(x^2+4x+8)/(x^2-4x+8)+\text{arc tan } (x+2)/2\right.$

$\left.+\text{arc tan } (x-2)/2\right\}/64+C.$

70. $I = -1/(x-1)^3+1/(x-1)^2-4/(x-1)+5 \log|x-1|+C.$

71. $I = \{\log (x^2+2x+5)/(x^2-2x+5)+\arctan (x+1)/2$
$+\arctan (x-1)/2\}/40+C.$

72. $I = -1/(x-1)^3-1/(x-1)^2-5/(x-1)+7 \log|x-1|$
$+2 \log|x+3|+C.$

73. $I = -1/2(x^2+4)+\frac{1}{2}\log|x^2+4|-\arctan\frac{1}{2}x+C.$

74. $I = -1/2(x-1)^2-5/2(x-1)+(3 \log|x-1|)/2$
$+(3 \log|x^2+1|)/4-\arctan x+C.$

EXERCISE 20, P. 33

1. $I=(2x+1)^{3/2}/3+C.$

2. $I=\frac{1}{2}\sqrt{(3+4x)}+C.$

3. $I=\frac{1}{2}(3x-4)^{2/3}+C.$

4. $I=5(2x+1)^{2/5}/4+C.$

5. $I=3(x^2-x+12)(x-4)^{1/3}/7+C.$

6. $I=(12x^2-x-1)(3x-1)^{1/3}/28+C.$

7. $I=2(2+3x)^{3/2}\{(2+3x)/5-2/3\}/9+C.$

8. $I=2(1-5x)^{3/2}\{(1-5x)/5-1/3\}/25+C.$

9. $I=3(x^2-x-12)(x-4)^{1/3}/7+C.$

10. $I=2(x-2)(2x+3)^{3/4}/7+C.$

11. $I=(5x^2-12x+36)(x+2)^{2/3}/40+C.$

12. $I=2(27x^2-12x+143)(3x+1)^{1/2}/405+C.$

13. $I=2(10x^2+3x-18)(2x+3)^{1/4}/45+C.$

14. $I=[\log|\{\sqrt{(x+a)}-\sqrt{a}\}/\{\sqrt{(x+a)}+\sqrt{a}\}|]/\sqrt{a}+C.$

15. $I=2\{\arctan\sqrt{(x/a-1)}\}/\sqrt{a}+C.$

16. $I=2\sqrt{x}+\log|(\sqrt{x}-1)/(\sqrt{x}+1)|+C.$

17. $I=2\sqrt{(1+x)}+\log|\{\sqrt{(x+1)}-1\}/\{\sqrt{(x+1)}+1\}|+C.$

18. $I = -(x+4\sqrt{x}+\log|1-\sqrt{x}|)+C.$

19. $I = \log\{\sqrt{(1-x)}-\sqrt{2}\}/\{\sqrt{(1-x)}+\sqrt{2}\}+C.$

20. $I = 4\sqrt{(1+\sqrt{x})}\{(1+\sqrt{x})^2/5-(1+\sqrt{x})/3\}+C.$

21. $I = 6\left(\frac{1}{2}t^2-t+\log|t+1|\right)+C,$ where $t = x^{1/6}$.

22. $I = 3\left(t^2-t+\frac{1}{2}\log|2t+1|\right)/2+C,$ where $t = x^{1/6}$.

23. $I = \{(x-5)^{3/2}-(x-7)^{3/2}\}/3+C.$ Rationalize the denominator.

24. $I = [\log|\{\sqrt{(x+9)}-3\}/\{\sqrt{(x+9)}+3\}|]/3+C.$

25. $I = 3\{(7-2x)^3/10-2(7-2x)^2+49(7-2x)/4\}(7-2x)^{1/3}/8 + C.$

26. $I = 2(x+1)^{1/2}-3(x+1)^{1/3}+6(x+1)^{1/6} - 6\log|1+(x+1)^{1/6}|+C.$

27. $I = 2\sqrt{\{(x-2)/(x-1)\}}+C.$

28. $I = \log|1-\sqrt{(1-x^2)}|-\log|x|-\text{arc sin } x+C.$

29. $I = -3(x+1)^{2/3}/2-6(x+1)^{5/6}/5-(x+1)-6(x+1)^{7/6}/7 - 3(x+1)^{4/3}/4-2(x+1)^{3/2}/3+C.$

30. $I = 3x^{4/3}/4-6x^{7/6}/7+x-6x^{5/6}/5+3x^{2/3}-2x^{1/2} + 6x^{1/3}-6x^{1/6}+3\log|x^{1/6}-1|+9\log|x^{1/6}+1|+C.$

EXERCISE 21, P. 48

1. $I = 2\sqrt{(4x^2+3x+1)}+C.$

2. $I = 5\sqrt{(36x^2-109x+77)}/18+C.$

3. $I = \text{arc sin } (x-1)+C.$

4. $I = \text{arc sin } \{(x+3)/4\}+C.$

5. $I = (\text{arc sin } 3x)/3+C.$

6. $I = \text{arc sin } (x/r-1)+C.$

7. $I = -\sqrt{(1-4x^2)}/4+3(\text{arc sin } 2x)/2+C.$

8. $I = -\sqrt{(1-2x-3x^2)}/3 - \{\text{arc sin}\,(3x+1)/2\}/3\sqrt{3}+C.$

9. $I = \frac{1}{2}x\sqrt{(1-4x^2)} + (\text{arc sin}\,2x)/4 + C.$

10. $I = -6\sqrt{(6+x-x^2)} + 8\{\text{arc sin}\,(2x-1)/5\}+C.$

11. $I = -\sqrt{(5+4x-x^2)} - 3\{\text{arc sin}\,(x-2)/3\}+C.$

12. $I = -\sqrt{(8+2x-x^2)} + 2\{\text{arc sin}\,(x-1)/3\}+C.$

13. $I = \frac{1}{2}(x-3)\sqrt{(6x-x^2)} - 9\{\text{arc sin}\,(3-x)/3\}/2+C.$

14. $I = 2\sqrt{(3-2x-x^2)} - 5\{\text{arc sin}\,(x+1)/2\}+C.$

15. $I = \log|x+3/2+\sqrt{(x^2+3x+2)}|+C.$

16. $I = \frac{1}{2}\log|2x+3/4+\sqrt{(4x^2+3x-1)}|+C.$

17. $I = \frac{1}{2}\log|x-1/2+\sqrt{(x^2-x+m)}|+C.$

18. $I = \log|x-2a+\sqrt{\{(x-a)(x-3a)\}}|+C.$

19. $I = \sqrt{(x^2+2x)} + 2\log|x+1+\sqrt{(x^2+2x)}|+C.$

20. $I = 3\sqrt{(x^2-5x+19)} + 19\{\log|x-5/2+\sqrt{(x^2-5x+19)}|\}$
$/2+C.$

21. $I = \sqrt{(x^2-ax)} + 3a\{\log|x-a/2+\sqrt{(x^2-ax)}|\}/2+C.$

22. $I = 3\sqrt{(4x^2-4x+5)}/4 - \{\log|2x-1+\sqrt{(4x^2-4x+5)}|\}$
$/4+C.$

23. $I = 3\sqrt{(x^2-4x+5)} + 8\log|x-2+\sqrt{(x^2-4x+5)}|+C.$

24. $I = 3\sqrt{(4x^2+5x-8)}/4 - 47\{\log|2x+5/4$
$+\sqrt{(4x^2+5x-8)}|\}/16+C.$

25. $I = 5\sqrt{(2x^2+8x-1)}/2 - 8\left\{\log\left|x+2+\sqrt{\left(x^2+4x-\frac{1}{2}\right)}\right|\right\}$
$/\sqrt{2}+C.$

26. $I = \frac{1}{2}(1+x)\sqrt{(4x+x^2)} - \frac{1}{2}\log|2+x+\sqrt{(2x+x^2)}|+C.$

27. $I = 5\sqrt{(3x^2-2x+1)}/3 - 7\{\log|x-1/3$
$+\sqrt{(x-2x/3+1/3)}|\}/3\sqrt{3}+C.$

28. $I = \frac{1}{2}(x+1)\sqrt{(3-2x-x^2)} + 2 \text{ arc sin } (x+1)/2 + C.$

29. $I = \frac{1}{2}x\sqrt{(x^2-4)} - 2 \log |x+\sqrt{(x^2-4)}| + C.$

30. $I = (3x+5)\sqrt{(3x^2+10x+9)}/6$
$+\{\log |3x+5+\sqrt{[3(3x^2+10x+9)]}|\}/3\sqrt{3} + C.$

31. $I = \frac{1}{2}(x-3/2)\sqrt{(x^2-3x+2)} - \{\log |x-3/2$
$+\sqrt{(x^2-3x+2)}|\}/8 + C.$

32. $I = -\frac{1}{2}x\sqrt{(1-x^2)} + \frac{1}{2} \text{ arc sin } x + C.$

33. $I = \frac{1}{2}(x-3)\sqrt{(x^2+2x+2)} + \frac{1}{2}\log |x+1+\sqrt{(x^2+2x+2)}|$
$+C.$

34. According to the substitution made we obtain one of the four following solutions. Substituting $\sqrt{x} = t$ we obtain arc sin $\sqrt{x} - \sqrt{(x-x^2)} + C_1$. Substituting $\sqrt{(1-x)} = t$ we obtain arc sin $\sqrt{(1-x)} - \sqrt{(x-x^2)} + C_2$. Substituting $\sqrt{\{x/(1-x)\}} = t$ we obtain arc tan $\sqrt{\{x/(1-x)\}} - \sqrt{(x-x^2)} + C_3$. Writing $\sqrt{\{x/(1-x)\}} = x/\sqrt{(x-x^2)}$ we obtain $\frac{1}{2}$ arc sin $(2x-1)$ $-\sqrt{(x-x^2)} + C_4$. The four constants are connected by the relationships $C_2 = C_1 + \pi/2$, $C_3 = C_1$, $C_4 = C_1 + \pi/4$.

35. $I = (x-3/a)\sqrt{(ax^2+2x+1)} + \{3 \log |ax+1$
$+\sqrt{[a(ax^2+2x+1)]}|\}/a\sqrt{a} + C.$

36. $I = (x+3)\sqrt{(x^2+1)} + C.$

37. $I = (x-a)\sqrt{(x^2+a^2)} + C.$

38. $I = (2x^2-5x+1)\sqrt{(x^2+2x+2)}/6$
$+\{5 \log |x+1+\sqrt{(x^2+2x+2)}|\}/2 + C.$

39. $I = (2x^2+x+7)\sqrt{(x^2+2x-1)}/6$
$-2 \log |x+1+\sqrt{(x^2+2x-1)}| + C.$

40. $I = (x^2+5x+24)\sqrt{(x^2-4x+3)}/3$
$+11 \log |x-2+\sqrt{(x^2-4x+3)}| + C.$

41. $I=(8x^2-10x-1)\surd(x^2+x+1)/8$
$+\{53 \log|x+1/2+\surd(x^2+x+1)|\}/16+C.$

42. $I=(3x^3-2x^2-10x-60)\surd(4x-x^2)/12+10 \text{ arc sin } (x-2)/2+C.$

43. $I=(8x^2-2x-51)\surd(6+x-x^2)/24$
$-\{25 \text{ arc sin } (1-2x)/5\}/16+C.$

44. $I=(5x^3-6x)\surd(5x^2+4)/100$
$+\{6 \log|x\surd 5+\surd(5x^2+4)|\}/25\surd 5+C.$

45. $I=(x^2/3+25x/12-163/24)\surd(x^2+x+1)$
$+\{85 \log|x+1/2+\surd(x^2+x+1)|\}/16+C.$

46. $I=(5x/6+17/12)\surd(3x^2-5x+8)$
$+\{55 \log|3x-5/2+\surd[3(3x^2-5x+8)]|\}/8\surd 3+C.$

47. $I=-(x^2/3+9x/2+227/6)\surd(5+6x-x^2)$
$-139 \text{ arc sin } (3-x)/\surd 14+C.$

48. $I=(x^2/3-x/12-67/24)\surd(8+x-x^2)$
$-\{33 \text{ arc sin } (1-2x)/\surd 33\}/16+C.$

49. $I=(2x^2/3-3x+1/6)\surd(2+3x-x^2)$
$-\{17 \text{ arc sin } (3-2x)/\surd 17\}/4+C.$

50. $I=(x^2-3)\surd(2x^2+3)/6+C.$

51. $I=(x^4-2x^2+6)\surd(2x^2+3)+C.$

52. $I=(x^3/4-7x^2/12+x/3+5/2)\surd(3+2x+x^2)$
$+\{7 \log|1+x+\surd(3+2x+x^2)|\}/2+C.$

53. $I=-\surd(10x-x^2)/5x+C.$

54. $x<-1,\ I=-\surd\{(x-1)/(x+1)\}+C,$
$x>1,\ I=\surd\{(x-1)/(x+1)\}+C.$

55. $I=-\frac{1}{2}\surd\{(2-x)/(2+x)\}+C.$

56. $x<-(\surd 5-1)/2,\ I=\text{arc sin } (2-x)/x\surd 5+C,$
$x>(\surd 5-1)/2,\ I=\text{arc sin } (x-2)/x\surd 5+C.$

57. $x<-\surd 2+1,\ I=\text{arc sin } (x+1)/x\surd 2+C,$
$x>\surd 2+1,\ I=-\text{arc sin } (x+1)/x\surd 2+C.$

58. $x<-1,\ I=\{\text{arc sin } (2-x)/(2x-1)\}/\surd 3+C,$
$x>1,\ I=-\{\text{arc sin } (2-x)/(2x-1)\}/\surd 3+C.$

59. $I = \frac{1}{2} \text{arc sin } 2x/(x+1) + C.$

60. $1 < x < 3/2,\ I = \{\text{arc sin } x/(3-2x)\}/\sqrt{3} + C,$
$3/2 < x < 3,\ I = \{\text{arc sin } x/(2x-3)\}/\sqrt{3} + C.$

61. $I = \log |x/\{x+2+2\sqrt{(x^2+x+1)}\}| + C.$

62. $x < -1,\ I = \text{arc sin } 1/x + C,$
$x > 1,\ I = -\text{arc sin } 1/x + C.$

63. $I = \sqrt{\{(a+x)/(a-x)\}}/a + C.$

64. $I = \{\text{arc sin } (x+5)/2\sqrt{2}(x-2)\}/\sqrt{7} + C.$

65. $I = -\sqrt{(x^2-4x+1)}/x + 2\log |\{1-2x-\sqrt{(1-4x+x^2)}\}/x| + C.$

66. $I = -\sqrt{(4-x^2)}/4x + C.$

67. $I = -\sqrt{(10x-x^2)}/9(x-1) + \{4 \log |[4x+5+3\sqrt{(10x-x^2)}]/(x-1)|\}/27 + C.$

68. $I = -\sqrt{(x^2+1)}/2x^2 - \frac{1}{2} \log |x/\{1+\sqrt{(x^2+1)}\}| + C.$

69. $I = (3x-1)\sqrt{(2x^2+2x+1)}/2x^2 - \frac{1}{2} \log |\{x+1+\sqrt{(2x^2+2x+1)}\}/x| + C.$

70. $I = (5-6x)\sqrt{(3-2x^2)}/2(x-1)^2 + 7 \log |\{3-2x-\sqrt{(3-2x^2)}\}/(x-1)| + C.$

71. $I = \sqrt{(1+x^2)}/2x^2 - \frac{1}{2} \log |\{1-\sqrt{(1+x^2)}\}/x| + C.$

72. $I = -(1/9x^3 + 5/54x^2 + 1/54x)\sqrt{(3-2x+x^2)} + 2 \log |\{3-x-\sqrt{[3(3-2x+x^2)]}\}/x|/27\sqrt{3} + C.$

73. $I = \{1/9(x-2)^3 + 2/27(x-2)\}\sqrt{(1-4x+x^2)} + C.$

EXERCISE 22, P. 67

1. $I = (\sin 12x)/24 + (\sin 2x)/4 + C.$

2. $I = -(\cos 5x)/5 - (\cos x)/2 + C.$

3. $I = (\sin 5x)/10 + (\sin x)/2 + C.$

4. $I = -(\cos 4x)/8 + (\cos 2x)/4 + C.$

5. $I = -(\cos 6x)/12 - (\sin 2x)/4 + C.$

6. $I = (\sin 3x)/6 - (\sin 7x)/14 + C.$

7. $I = (\sin 4x)/8 + (\sin 2x)/4 + C.$

8. $I = (\sin 2x)/4 - (\sin 4x)/8 + C.$

9. $I = (\sin 3x)/6 - (\sin 7x)/14 + C.$

10. $I = (\cos^3 x)/3 - \cos x + C.$

11. $I = -\cos x \sin x\,(\cos^2 x + 3/2)/4 + 3x/8 + C.$

12. $I = \cos x \sin x\,(\cos^2 x + 3/2)/4 + 3x/8 + C.$

13. $I = (\sin^5 x)/5 - (2 \sin^3 x)/3 + \sin x + C.$

14. $I = -(\cos^5 x)/5 + (2 \cos^3 x)/3 - \cos x + C.$

15. $I = (\tan^4 x)/4 - (\tan^2 x)/2 - \log \cos x + C.$

16. $I = -(\cot^3 x)/3 + \cot x + x + C.$

17. $I = -(\cot^5 x)/5 + (\cot^3 x)/3 - \cot x - x + C.$

18. $I = (\sec^6 x \sin x)/6 + (5 \sec^4 x \sin x)/24$
$+ (15 \sec^2 x \sin x)/48 + (15 \log |\sec x + \tan x|)/48.$

19. $I = -(\operatorname{cosec}^5 x \cos x)/5 - (4 \operatorname{cosec}^3 x \cos x)/15$
$- (8 \cot x)/15 + C.$
$= -\cot x - (2 \cot^3 x)/3 - (\cot^5 x)/5 + C.$

20. $I = (\cos^7 x)/7 - (\cos^5 x)/5 + C.$

21. $I = (\cos^{13} x)/13 - (3 \cos^{11} x)/11 + (\cos^9 x)/3 - (\cos^7 x)/7 + C.$

22. $I = \{\sin^6 x - (\sin^4 x)/5 - (4 \sin^2 x)/15 - 8/15\} (\cos x)/7 + C.$

23. $I = (\sin^3 x \cos x)/4 - (\sin x \cos x)/8 + x/8 + C.$

24. $I = (\sin^4 x)/4 - (\sin^6 x)/6 + C.$

25. $I = (\sin^5 x)/5 - (2 \sin^7 x)/7 + (\sin^9 x)/9 + C.$

26. $I = -(\operatorname{cosec}^7 x)/7 + C.$

27. $I = \log|\tan(\pi/4 + x/2)| - \sin x + C.$

28. $I = 3(\sin x)^{1/3} + C.$

29. $I = -3(1 + 2\cos x)^{2/3}/4 + C.$

30. $I = -2(1 + \cos^2 x)^{1/2} + C.$

31. $I = \log|1 + \sin^2 x| + C.$

32. $I = \arcsin(\sin^2 x) + C.$

33. $I = -(1 + \sin^2 x)/\sin x + C.$

34. $I = \sin x - \cos x + C.$

35. $I = (\sin x \sec^4 x)/4 + (3 \sin x \sec^2 x)/8$
$+ (3 \log|\tan(\pi/4 + x/2)|)/8 + C.$

36. $I = -(\operatorname{cosec}^6 x)/6 + (5 \operatorname{cosec}^4 x)/24 + (5 \operatorname{cosec}^2 x)/16 \cos x$
$+ \left(5 \log\left|\tan \frac{1}{2}x\right|\right)/16 + C.$

37. $I = -\frac{1}{2} \operatorname{cosec}^2 x + \log|\tan x| + C.$

38. $I = \frac{1}{2} \tan^2 x + \log|\tan x| + C.$

39. $I = \frac{1}{2} \sec^2 x - \operatorname{cosec}^2 x - (\operatorname{cosec}^4 x)/4 + 3 \log|\tan x| + C.$

40. $I = (\sec^3 x \operatorname{cosec} x)/3 - (8 \cot 2x)/3 + C.$

41. $I = \sin x + \frac{1}{2} \sin x \sec^2 x + (3 \log|\tan(\pi/4 - x/2)|)/2 + C.$

42. $I = -(\sin^3 x)/3 - \sin x + \log|\tan(\pi/4 + x/2)| + C.$

43. $I = \frac{1}{2}(\sin^2 x - \operatorname{cosec}^2 x) - 2 \log|\sin x| + C.$

44. $I = (\sec^7 x)/7 - (\sec^5 x)/5 + C.$

45. $I = -\frac{1}{2} \sin x \sec^2 x + (3 \log|\tan(\pi/4 + x/2)|)/2 + C.$

46. $I = \{2 \arctan[(\tan x/2)/3]\}/3 + C.$

47. $I = \cot(\pi/4 + x/2) + C$.

48. $I = (\log|\tan(\pi/8 + x/2)|)/\sqrt{2} + C$. Use the transformation $\sin x + \cos x = \sin x + \sin(\pi/2 + x)$
$= 2\sin(\pi/4 + x)\cos(-\pi/4) = \sqrt{2}\sin(\pi/4 + x)$.

49. For $\cos x \neq 0$, $I = \frac{1}{2}\arctan(\tan^2 x) + C_1$, and

for $\sin x \neq 0$, $I = -\frac{1}{2}\arctan(\cot^2 x) + C_2$.

50. $I = 2\tan x + \{\arctan(\sqrt{2}\tan x)\}/\sqrt{2} + C$.

51. $I = -1/(\sin x - \cos x) + C$.

52. $I = \{\log(\sqrt{2} - \sin 2x)/(\sqrt{2} + \sin 2x)\}/2\sqrt{2} + C$.

53. $I = \frac{1}{2}\arctan(2\tan^2 x + 1) + C$.

54. $I = \{3\arctan(\tan x/\sqrt{2})\}/4\sqrt{2} - \tan x/4(\tan^2 x + 2) + C$.

55. $I = \{\sqrt{(2+\sqrt{2})}\arctan[(\tan 2x)/\sqrt{(4+2\sqrt{2})}]$
$-\sqrt{(2-\sqrt{2})}\arctan[(\tan 2x)/\sqrt{(4-2\sqrt{2})}]\}/4 + C$.

56. $I = \{\arctan[(\tan 2x)/\sqrt{2}]\}/\sqrt{2} + C$.

57. $I = \frac{1}{2}\tan x + \{\arctan(\sqrt{2}\tan x)\}/2\sqrt{2} + C$.

EXERCISE 23, P. 70

1. $I = x^2/4 - \frac{1}{2}x\sqrt{(1-x^2)}\arcsin x + (\arcsin x)^2/4 + C$.

2. $I = (x\arcsin x)/\sqrt{(1-x^2)} + \frac{1}{2}\log|1-x^2| + C$.

3. $I = x\arctan x - \frac{1}{2}\log|1+x^2| - \frac{1}{2}(\arctan x)^2 + C$.

4. $I = \{2\sqrt{(\arctan 3x)}\}/3 + C$.

5. $I = 1/2\ \mathrm{arc\,cot}\ 2x + C$.

6. $I = (\arctan x)^3/3 + C$.

7. $I = 1/\text{arc cos } x + C.$

8. $I = \log |\text{arc sin } x| + C.$

9. $I = \{(x^2-1) \text{ arc tan } x + x\}/4(1+x^2) + C.$

10. $I = \text{arc sin } x/\sqrt{(1-x^2)} + \frac{1}{2} \log |(1-x)/(1+x)| + C.$

11. $I = \frac{1}{2}\left(x^2 - \frac{1}{2}\right) \text{arc sin } x + \{x\sqrt{(1-x^2)}\}/4 + C.$

12. $I = \log |\{(x-1)/(x+1)\}^{1/8}| - \frac{1}{2} \text{ arc tan } x \left\{\frac{1}{2} + 1/(x^2-1)\right\}$
$+ C.$

13. $I = (x^3 \text{ arc tan } x)/3 - x^2/6 + (\log |1+x^2|)/6 + C.$

14. $I = x - \log(1+e^x) - 2e^{-x/2} \text{ arc tan } e^{x/2} - \left(\text{arc tan} \frac{1}{2} x\right)^2 + C.$

15. $I = -(\text{arc sin } x)/x - \log |\{1+\sqrt{(1-x^2)}\}/x| + C.$

16. $I = x - e^x \text{ arc sin } (e^x) - \log \{1+\sqrt{(1+e^{2x})}\} + C.$

17. $I = \left(x^7 \text{ arc tan } x - x^6/6 + x^4/4 - x^2/2 + \frac{1}{2} \log|1+x^2|\right)/7 + C.$

18. $I = -(2x+21)\sqrt{(-x^2+3x-2)}/4$
$+ (x^2+3x-55/8) \text{ arc cos } (2x-3) + C.$

19. $I = \sqrt{(1+x^2)} \text{ arc tan } x - \log |x + \sqrt{(1+x^2)}| + C.$

20. $I = -x^2/8 + \{x\sqrt{(1-x^2)} \text{ arc sin } x\}/4 + (\text{arc sin } x)^2/8 + C.$

21. $I = x/4 + x^3/12 + \{(1+x^2)^2 \text{ arc tan } x\}/4 + C.$

22. $I = -2 \sin (1-x)\sqrt{x} + (1+x) \text{ arc sin } \{2\sqrt{x}/(1+x)\} + C.$

EXERCISE 24, P. 73

1. $I = e^{3x}/3 + 2\sqrt{(e^x)} + C.$

2. $I = \log(e^x + e^{-x}) + C = \log(2 \cosh x) + C.$

3. $I = \frac{1}{2} \log |e^{2x} - 1| - x + C.$

4. $I = \text{arc tan } (e^x) + C.$

5. $I = 2\sqrt{(e^x+1)} + \log\{\sqrt{(e^x+1)}-1\}/\{\sqrt{(e^x+1)}+1\} + C.$

6. $I = 2\log|e^x+1| - x + C.$

7. $I = [\log\{\sqrt{(3+2e^x)}-\sqrt{3}\}/\sqrt{(3+2e^x)}+\sqrt{3}\}]/\sqrt{3} + C.$

8. $I = 2(1+e^x)^{3/2}/3 + C.$

9. $I = -(e^x-1)^{-1} + C.$

10. $I = \frac{1}{2}(e^{2x}-e^{-2x}) + 2x + C = 2\sinh x\cosh x + 2x + C.$

11. $I = \log|e^x+5| + C.$

12. $I = (35\log|e^{2x}-4/9|)/36 - 3x/2 + C.$

13. $I = \log|1+e^x| - e^{-x} - x + C.$

14. $I = -1/(n-1)(e^x+a)^{n-1} + C.$

15. $I = \{\text{arc sin}\sqrt{(5/3)e^x}\}/\sqrt{5} + C.$

16. $I = x - \log|2e^x+1| + \sqrt{(e^{2x}+4e^x+1)} + C.$

17. $I = -e^{-x}(x^3+3x^2+6x+6) + C.$

18. $I = \log|\log x| + C.$

19. $I = x\log|x^2+1| - 2x + 2\,\text{arc tan}\,x + C.$

20. $I = x(\log|x|)^2 - 2x\log|x| + 2x + C.$

21. $I = x\log|x+\sqrt{(x^2+1)}| - \sqrt{(x^2+1)} + C.$

22. $I = (2+5x)(\log|2+5x|-1)/5 + C.$

23. $I = \text{arc tan}\log|x| + C.$

24. $I = -(\log|x|+1)/x + C.$

25. $I = (4+3x)^3(\log|x|)/9 - (64\log|x|)/9 - 16x - 6x^2 - x^3 + C.$

26. $I = x^4(\log|x^2+3|)/4 - (x^4/2 - 3x^2$
$+ 9\log|x^2+3|)/4 + C.$

27. $I = a^x(x - 1/\log a)/\log a + C.$

EXERCISE 25, P. 86

1. $\frac{1}{2}\log(21/5).$ **2.** $\pi/4a.$

3. $1/2.$ **4.** $\pi/20.$

5. $(3\log 1/9)/8$. **6.** $\pi/6$.

7. $1-\sqrt{3}/2-\pi/6$. **8.** $\pi/\sqrt{5}$.

9. $\frac{1}{2}(15^{2/3}-5^{2/3})$. **10.** $\frac{1}{2}\log(18+\sqrt{10})$, put $x^2 = u$.

11. $(5\sqrt{5}-8)a^3$.

12. $3\sqrt{7}/2-\sqrt{2}+\log(2+\sqrt{2})/(3+\sqrt{7})$.

13. $\log 5(3+2\sqrt{3})/(6+\sqrt{111})$.

14. $\pi a^2/4$. **15.** $\pi/12-1/6+(\log 2)/6$.

16. $e^2-1+\log 2/(1+e^2)$.

17. $-2/e+1$. **18.** $2e^4-e/2$.

19. $e^2(5e^2-1)/4$, **20.** $(3e^{\pi}-1)/8$.

21. 1. **22.** $\log 3$.

23. $4/3$.

24. $(4\arctan 3/\sqrt{3})/3\sqrt{3}-1/6-(4\arctan 1/\sqrt{3})/3\sqrt{3}$.

25. $1-\{\log(\sqrt{2}+1)/(\sqrt{2}-1)\}/2\sqrt{2}-\sqrt{(2/3)}$
$+\{\log(\sqrt{2}+\sqrt{(3/2)})/(\sqrt{2}-\sqrt{(3/2)})\}/2\sqrt{2}$.

26. $\{\arctan(1/5)-\arctan(1/10)\}/5$.

27. $-1/2+\log\sqrt{3}$. **28.** $\pi/2$.

29. $2(\sqrt{2}-1)$. **30.** $2(\sqrt{2}-1)$.

31. $\pi/4+\log(\sqrt{2}/2)$. **32.** $1/336$.

33. $5\pi/1024$, the function is even, thus $\int_0^{\pi} = 2\int_0^{\pi/2}$.

34. $9\sqrt{2}/560$.

35. $a^2/3$. This result was known to Archimedes.

36. $1/3$. **37.** $8/3$.

38. 8. **39.** $5/6$.

40. $5/12$. **41.** $343/3$.

42. $64/3$. **43.** 0·950.

44. $32/3$. **45.** $27/2$.

46. $9/2$. **47.** $(3\pi+2)/(9\pi-2)$.

48. $15/2-8 \log 2$.

49. $6(3\pi+8)$.

50. $1/16$.

51. $1/4-1/2e$.

52. $3\pi(\sqrt{3}-1/2)/2+9(1-\sqrt{3})/2$.

EXERCISE 26, P. 95

1. $\pi a^2/4$. The curve is a circle of radius a and centre $(a/2, \pi/2)$.

2. When θ lies in either of the intervals $\pi/2 \leqslant \theta \leqslant \pi$, and $3\pi/2 \leqslant \theta \leqslant 2\pi$, we obtain a negative value for r. If $r_0 > 0$, then by the point $(-r_0, \theta_0)$ we understand the point $(r_0, \theta_0+\pi)$. Under this convention the equation represents a four-leaved rosette. Area $= \pi a^2/2$.

3. $\pi a^2/2$. The curve is a two-leaved rosette.

4. $3\pi a^2/2$. The curve is symmetrical about the initial line.

5. $4\pi^3 a^2/3$. Draw a rough graph.

6. πa^2.

7. $\pi p^2/(1-e^2)^{3/2}$.

8. $72\sqrt{3}/5$.

9. $3a^2/2$. Draw a rough graph considering the intervals $0 < t < t_0$, and $t_0 < t < \infty$.

10. $a^2(\pi/4+1-\sqrt{3})$.

11. $4\sqrt{3}/5$.

12. $8/15$.

13. $3(4\pi^3+3\pi)$.

14. 24π.

EXERCISE 27, P. 99

1. $52/57$.

2. $14/3$.

3. $4(2\sqrt{2}-1)/3$.

4. $14/9$.

5. $56/3$.

6. $16(31\sqrt{31/8}-1)/81$.

7. $16(13\sqrt{13/8}-1)/27$.

8. $3\sqrt{10}-\sqrt{2}+\dfrac{1}{2}\log(3+3\sqrt{10})/(3/2+\sqrt{2})$.

9. $\pi/2$.

10. $\pi/4$.

11. $\sqrt{2}+\log(1+\sqrt{2})$. We take y as the independent variable.

12. $2+(3\log 3)/2$.

13. $14/3$.

14. $(\log 3)/2$.

15. $\log(\sqrt{2}+1)$.

16. $a(e-e^{-1})$.

17. $1+\frac{1}{2}\log 3/2$.

18. $\log 3-1/2$.

19. $\log(e^{1/2}+e^{-1/2})$.

20. π.

21. $a\{2+\sqrt{3}\log(2+\sqrt{3})\}$.

22. $3/4+\frac{1}{2}\log 2$.

23. $\sqrt{(1+e^2)}-\sqrt{2}+\log\{\sqrt{(1+e^2)}-1\}-1-\log(\sqrt{2}-1)$.

24. $4a$.

25. $a\{\sqrt{2}+\log(1+\sqrt{2})\}/2$.

26. $5/12+\log 3/2$.

27. $2\pi a$.

28. $p\{\sqrt{2}+\log(1+\sqrt{2})\}$.

29. $2\sqrt{3}$.

30. $\pi^2 a/2$.

31. $5a\{2+[\log(2+\sqrt{3})]/\sqrt{3}\}/8$.

32. $2a\left\{7/2-\sqrt{3}\log\frac{1}{2}(\sqrt{3}+\sqrt{7})\right\}$.

33. 40.

34. $a\{1+[\log(\sqrt{2}+1)/(\sqrt{2}-1)]/2\sqrt{2}\}/2$.

35. 10π.

EXERCISE 28, P. 105

1. $V=\pi\log(b/a)$.

2. $V=\pi$.

3. $V=\pi^2/2,\quad S=2\pi\{\sqrt{2}+\log(\sqrt{2}+1)\}$.

4. $V=\pi(15/2-8\log 2)$.

5. $V=\pi a^3(2-\sqrt{2})/3$,

$S=\pi a^2(\sqrt{6}-1)-a^2\{\log(2+\sqrt{3})/(1+\sqrt{2})\}/4$.

6. $V=\pi/63,\quad S=2\pi\sqrt{2}/9$.

7. $V=\pi c^3(e^2+4-e^{-2})/4,\quad S=\pi c^2(e^2+4-e^{-2})/2$.

8. $S=4\pi r$

9. $V=\pi\{7/3+3a+4a^2+8a^3\log|(2a-2)/(2a-1)|\}$,

$$S=2\pi a\big[(x-4a)\sqrt{(8ax-3x^2)}/(x-2a)+8a\{\arccos(3x-4a)/4a\}/3\big]_1^2.$$

10. $S = a\sqrt{\{b^2+(2\pi a)^2\}}/2+\{b^2 \log|2\pi a/b + \sqrt{(4\pi^2a^2/b^2+1)}|\}/4\pi.$

11. $16\pi/35.$

12. $V=\pi\{3\sqrt{10}+10\log|\sqrt{10}-2|-3\sqrt{15}-10\log|\sqrt{15}-3|\}.$

13. $V=16\pi.$

14. $S=56\pi/27.$

15. $V=64\pi.$

16. $S=37\pi\sqrt{37}/54.$

17. $V=16\pi/35.$

18. $V=2\pi/3.$

19. $V=\dfrac{1}{2}\pi\log 9/5.$

20. $V=200\pi/3.$

21. $S=200\pi^2, \quad V=500\pi^2.$

22. $V=\pi/5(1-e^{-2\pi}).$

23. $V=64\pi a^3/315, \quad S=16\pi a^2/35.$

24. $V=3\pi/4, \quad S=3\pi.$

EXERCISE 29, P. 117

1. $Mb^2/3$, where M is the mass of the rectangle.

2. $Mr^2/2.$

3. $Mh^2/6.$

4. $3Mr^2/10.$

5. $Mr^2/2.$

6. $Mr^2/4.$

7. $2Mr^2/5.$

8. $2Mab/5.$

9. $mc^3(e+e^{-1})(e^2+10+e^{-2})$, where m is the mass per unit length.

10. $\bar{y}=2r/\pi.$

11. $\bar{y}=4r/3\pi.$

12. $\bar{x}=3r/8.$

13. $h/4$ from the base.

14. $\bar{x}=h/3.$

15. $c(e^2+4-e^{-2})/2(e-e^{-1}).$

16. $\bar{y}=5a/6.$

17. $\bar{x}=4a/5.$

18. $\bar{x}=r^3(92-45\sqrt{3}-10\pi)/24A,$
$\bar{y}=8r^3\{(3\sqrt{3}-7)/24+\log(\sqrt{3}/2\cos\pi/12)\}/A$, where A is the area.

19. $\bar{x}=9/7, \quad \bar{y}=5\sqrt{3}/32.$

20. $\bar{x}=7/5, \quad \bar{y}=\sqrt{3}/4.$

21. $\bar{x}=a(\pi/2+64/45\pi).$

22. $\bar{x}=256a/315\pi, \quad \bar{y}=256a/315\pi.$

23. $\bar{x}=b(\pi-2)/4\pi, \quad \bar{y}=\pi a/8.$

24. $\bar{x}=83/77, \quad \bar{y}=9/154.$

25. $\bar{x}=9a/20, \quad \bar{y}=9a/20.$

26. $\bar{r}=5a/6, \quad \bar{\theta}=0.$

27. $\bar{x}=3x_0/5, \quad \bar{y}=3y_0/5.$

EXERCISE 30, P. 126

1. $+\infty$

2. $3/2.$

3. $+\infty$

4. $8.$

5. $4\sqrt{2}.$

6. $5/2.$

7. $\pi/2.$

8. $\pi.$

9. $\pi/12.$

10. $1.$

11. $3\sqrt[3]{2}/8.$

12. $\log(2+\sqrt{3}).$

13. $\sqrt{5}.$

14. $9\sqrt[3]{4}.$

15. $3\pi/8.$

16. $\pi/3.$

17. $-\pi/3.$

18. $\pi/2.$

19. $\pi/2.$

20. $7\pi/2.$

21. $\log 3.$

22. ∞, the integral is not convergent.

23. $\infty.$

24. $\pi.$

25. $\pi^3/12.$

26. $\pi/9.$

27. $1/3.$

28. $1.$

29. $\sqrt{2}\pi/8-\sqrt{2}\{\text{arc tan}\sqrt{2}/2\}/4.$

30. $3\pi/8.$

31. $\pi/3.$

32. $\pi.$

33. $-\frac{1}{2}+\log 2.$

34. $2\pi/3\sqrt{3}.$

35. $\pi/2\sqrt{3}-\frac{1}{2}\log 25/12.$

36. $\pi/2$.

37. $1/a$.

38. $1/2$.

39. $1/2$.

40. $e-1$.

41. $\pi/\sqrt{(a^2-b^2)}$.

42. $\pi/2ab$.

EXERCISE 31, P. 132

1. The function has no extreme values.

2. Minimum at the points $(-\sqrt{2}, \sqrt{2})$ and $(\sqrt{2}, -\sqrt{2})$, $z_{min} = -8$. The point (0, 0) is not an extreme point.

3. Minimum at (0, 1), $z_{min} = -7$.

4. There is an extreme at the point (a, a), where $z = -a^3$. For $a > 0$ the function is a minimum, for $a < 0$ the function is a maximum.

5. The first partial derivatives are zero at the points $A(0, -5/2)$, $B(-1, -1)$ and $C(3, -1)$, but only at A is there an extreme value, when $z_{max} = 75/4$.

6. Minimum at $(-4/3, 1/3)$, $z_{min} = -4/3$.

7. Minimum at $(8/3, 2/3)$, $z_{min} = -11/3$.

8. Minimum at $(27/2, 5)$, $z_{min} = -109/4$.

9. Maximum at (a, a), $z_{max} = -a^3$.

10. Minimum at $\left(\frac{1}{2}\sqrt[3]{2}, \frac{1}{2}\sqrt{2}\right)$, $z_{min} = 3/\sqrt[3]{2}$.

11. Minimum at (1, 0), $z_{min} = -1$.

12. Minimum at (8, 24), $z_{min} = -448$.

13. Minimum at (−1, 3), $z_{min} = -7$.

14. Maximum at (0, 0), $z_{max} = 1$.

15. Maximum at $(a/6, a/3)$, $z_{max} = a^6/432$.

16. Maximum at (2, 3), $z_{max} = 108$.

17. Maximum at $(\pi/3, \pi/3)$, $z_{max} = 3\sqrt{3}/2$.

18. Maximum at $(\pi/3, \pi/6)$, $z_{max} = 3\sqrt{3}/2$.

19. Maximum at $(a/3, a/3)$, $z_{max} = \sin^3(a/3)$.

20. Maximum at $(\pi/3, \pi/3)$, $z_{max} = 3\sqrt{3}/8$;
minimum at $(2\pi/3, 2\pi/3)$, $z_{min} = -3\sqrt{3}/8$.

21. The point (1, 1) where $z = -1+\frac{1}{2}\log 2+3/4$, is a saddle point, the three-dimensional equivalent of a point of inflexion.

22. Minimum at (1, 2), $z_{min} = 7-10\log 2$.

23. Minimum at (0, 0), $z_{min} = 0$; saddle point at $(-1/4, -1/2)$.

24. Maximum at (1, 3), $z_{max} = e^{-13}$; minimum at $(-1/26, -3/26)$, $z_{min} = -26e^{-1/52}$.

25. If the angles subtended at the centre of the circle by the three sides are x, y and $2\pi-(x+y)$, we obtain $A = \frac{1}{2}r^2\{\sin x+\sin y+\sin(2\pi-x-y)\}$, where r is the radius of the circle. The area A is a maximum when the angles are equal.

26. The number is divided into three equal parts.

27. The area is a maximum when the triangle is equilateral.

28. $P_1(3, -3)$, $P_2(9, 5)$.

29. (2, 2).

30. A cube of edge $2\sqrt{3}r/3$.

EXERCISE 32, P. 137

1. $y = 2x, \quad x+2y-5 = 0.$

2. $y = x-1, \quad y+x = 1.$

3. $2y = x, \quad x+2y-5 = 0.$

4. $2x+y-4 = 0, \quad x-2y+3 = 0.$

5. $x-y\sqrt{3}+5/3 = 0, \quad 3x+y\sqrt{3}-17/3 = 0.$

6. $3x-2y-p = 0,$

7. $2x-y-r = 0.$

8. $x+y-3a = 0.$

9. $y = a\log ax+a(1-\log a).$

10. $29x+27y-6 = 0.$

11. $(e^2-1)x-2ey+2c = 0.$

13. $6x-5y+21 = 0.$

14. $x = 2\pi a.$

15. k odd, $y = x + e^{k\pi}$, $y = -x - e^{k\pi}$;

k even, $y = x - e^{k\pi}$, $y = -x + e^{k\pi}$.

16. $y = x - \pi r\sqrt{2}/4$, $x + y = r\sqrt{2}$.

EXERCISE 33, P. 143

1. $\sqrt{2/3}$.

2. p.

3. $17^{3/2}/2$.

4. 8.

5. $3\sqrt{3}/64$.

6. $\varkappa(0) = b/a^2$, $\varkappa(a) = a/b^2$.

7. Where $x = \pm 1$, $|\varkappa_{\max}| = 1/\sqrt{2}$.

8. $(a^4+1)^{3/2}/2a^4$.

9. 36.

10. $108^{3/2}/6$.

11. $3m/8$, $3m/2$.

12. $5^{3/2}/4$.

13. $a \sec (x/a)$.

14. $-1/12$.

15. $3\sqrt{6}$.

16. $37^{3/2}/18$.

17. $(1+4e^2)^{3/2}/3e$.

18. c.

19. 1.

20. at_0.

21. $13^{3/2}/6$.

22. $2a$.

23. $R = 3a \cos^2 (\theta/3)/4$. Where $\theta = 0$ or $\theta = 3\pi$, $x = a$ and $y = 0$, and $R_{\max} = 3a/4$; where $\theta = 3\pi/2$, $x = 0$, $y = 0$, and $R_{\min} = 0$.

24. a.

25. $a(1+\theta_0^2)^{3/2}/(2+\theta_0^2)$.

26. $a/3$.

27. $\sqrt{2}mv^2/4$.

28. $|a|$.

EXERCISE 34, P. 151

1. $(x-15)^2+(y-6)^2 = 298$.

2. $x^2+y^2 = 1$, $x^2+(y-2)^2 = 1$.

3. $(x-2a/3)^2+(y-a/3)^2 = 8a^2/9$.

4. $27py^2 = 8(x-p)^3$, a semi-cubical parabola.

5. $(ax)^{2/3}+(by)^{2/3} = (a^2-b^2)^{2/3}$.

6. $(ax)^{2/3}-(by)^{2/3} = (a^2+b^2)^{2/3}$.

7. $(x+y)^{2/3}-(x-y)^{2/3} = (4a)^{2/3}$.

8. $x = 3(1+2t^2-t^4)/2,\ y = -4t^3$.

9. $x = a \log\{y \pm \sqrt{(y^2-4a^2)}\}/2a \mp y\sqrt{(y^2-4a^2)}/4a$.

10. $x = a\,(\text{arc cos } t - t^{-1}\sqrt{(1-t^2)}),\quad y = a\,(\log t - 1)$.

11. $x = a(t-\sin t)+\pi a,\quad y = a(1-\cos t)-2a$, which is the same cycloid translated.

12. $x = a \log \tan \frac{1}{2} t,\ y = a \operatorname{cosec} t = \frac{1}{2} a\left(\tan \frac{1}{2} t + \cot \frac{1}{2} t\right)$.

By making the substitution $\tan \frac{1}{2} t = e^{x/a}$ we obtain the equation of the evolute as the catenary $y = a \cosh (x/a)$.

13. $x = 3a(2 \cos t - \cos 2t),\ y = 3a(2 \sin t + \sin 2t)$, a hypocycloid.

EXERCISE 35, P. 162

1. $2\{\sin x - (\sin 2x)/2 + (\sin 3x)/3 - \ldots + (-1)^{n+1} (\sin nx)/n + \ldots\}$.

For $x = \pm\pi$ the series is zero.

2. $4a\{\sin x + (\sin 3x)/3 + (\sin 5x)/5 + \ldots\}/\pi$.

3. $4a\{\sin x + (\sin 3x)/3 + (\sin 5x)/5 + \ldots\}/\pi$. For $x = \pm\pi$ the Fourier series gives the value zero.

4. $4a\{\cos\alpha \sin x + (\cos 3\alpha \sin 3x)/3 + (\cos 5\alpha \sin 5x)/5 + \ldots\}/\pi$.

5. $\pi - 8\{\cos x + (\cos 3x)/3^2 + (\cos 5x)/5^2 + \ldots\}/\pi + 8\{\sin x - (\sin 2x)/2 + (\sin 3x)/3 - (-1)^{n+1}(\sin nx)/n + \ldots\}$.

For $x = \pm\pi$ the Fourier series takes the value 2π.

6. $4a\{\sin\alpha \sin x + (\sin 3\alpha \sin 3x)/3^2 + (\sin 5\alpha \sin 5x)/5^2 + \ldots\}/\pi\alpha$.

8. As a series of sines:
$f(x) = 3\{\sin x + (\sin 3x)/3^3 + (\sin 5x)/5^3 + \ldots\}/\pi$.

As a series of cosines:
$f(x) = \pi^2/6 - \{\cos 2x + (\cos 4x)/2^2 + (\cos 6x)/3^2 + \ldots\}$.

9. $2/\pi - 4\{(\cos 2x)/1.2 + (\cos 4x)/3.5 + (\cos 6x)/5.7 + \ldots\}/\pi$.

10. $\pi^2/3-4\{\cos x-(\cos 2x)/2^2+(\cos 3x)/3^2-\ldots\}$.

11. $1/\pi-2\{(\cos 2x)/1.3+(\cos 4x)/3.5+(\cos 6x)/5.7$
$+\ldots\}/\pi+(\sin x)/2$.

12. $\sum_{n=1}^{\infty}(-1)^n(1/n^3-2\pi^2/n)\sin nx$.

13. $4\{(2\sin 2x)/1.3+(4\sin 4x)/3.5+(6\sin 6x)/5.7+\ldots\}/\pi$.

14. $2\sin a\pi\{(\sin x)/(1-a^2)-(2\sin 2x)/(4-a^2)+\ldots$
$+(-1)^{n+1}(n\sin nx)/(n^2-a^2)+\ldots\}/\pi$.

15. $(\sin x)/2+\sum_{n=2}^{\infty}(-1)^{n+1}(n\sin nx)/(2n-3)(2n-1)$.

16. $\cos x+(\cos 3x)/3+(\cos 5x)/5+\ldots$.

17. $\cos x-(\cos 2x)/2+(\cos 3x)/3-(\cos 4x)/4+\ldots$.

18. $\cos x+(\cos 2x)/2+(\cos 3x)/3+\ldots$.

19. $2\alpha\sin\alpha x\{1/2\alpha^2-(\cos x)/(\alpha^2-1)+(\cos 2x)/(\alpha^2-4)-\ldots$
$+(-1)^n(\cos nx)/(\alpha^2-n^2)+\ldots\}/\pi$.

20. $4\{(2\sin 2x)/1.3+(4\sin 4x)/3.5+(6\sin 6x)/5.7+\ldots\}/\pi$.

21. $1/\pi+(\sin x)/2-(2/\pi)\sum_{n=1}^{\infty}(\cos 2nx)/(4n^2-1)$.

22. $(4B_0/\pi)\sum_{n=0}^{\infty}(\sin(2n+1)x)/(2n+1)$.

INDEX

MADE IN GREAT BRITAIN

Printed in Dunstable, United Kingdom

91338439R00118